T0260715

6LoWPAN

WILEY SERIES IN COMMUNICATIONS NETWORKING
& DISTRIBUTED SYSTEMS

Series Editor: David Hutchison, *Lancaster University, Lancaster, UK*
Serge Fdida, *Université Pierre et Marie Curie, Paris, France*
Joe Sventek, *University of Glasgow, Glasgow, UK*

The 'Wiley Series in Communications Networking & Distributed Systems' is a series of expert-level, technically detailed books covering cutting-edge research, and brand new developments as well as tutorial-style treatments in networking, middleware and software technologies for communications and distributed systems. The books will provide timely and reliable information about the state-of-the-art to researchers, advanced students and development engineers in the Telecommunications and the Computing sectors.

Titles in the series:

Wright: *Voice over Packet Networks* 0-471-49516-6 (February 2001)
Jepsen: *Java for Telecommunications* 0-471-49826-2 (July 2001)
Sutton: *Secure Communications* 0-471-49904-8 (December 2001)
Stajano: *Security for Ubiquitous Computing* 0-470-84493-0 (February 2002)
Martin-Flatin: *Web-Based Management of IP Networks and Systems* 0-471-48702-3 (September 2002)
Berman, Fox, Hey: *Grid Computing. Making the Global Infrastructure a Reality* 0-470-85319-0 (March 2003)
Turner, Magill, Marples: *Service Provision. Technologies for Next Generation Communications* 0-470-85066-3 (April 2004)
Welzl: *Network Congestion Control: Managing Internet Traffic* 0-470-02528-X (July 2005)
Raz, Juhola, Serrat-Fernandez, Galis: *Fast and Efficient Context-Aware Services* 0-470-01668-X (April 2006)
Heckmann: *The Competitive Internet Service Provider* 0-470-01293-5 (April 2006)
Dressler: *Self-Organization in Sensor and Actor Networks* 0-470-02820-3 (November 2007)
Berndt: *Towards 4G Technologies: Services with Initiative* 0-470-01031-2 (March 2008)
Jacquenet, Bourdon, Boucadair: *Service Automation and Dynamic Provisioning Techniques in IP/MPLS Environments* 0-470-01829-1 (March 2008)
Minei/Lucek: *MPLS-Enabled Applications: Emerging Developments and New Technologies, Second Edition* 0-470-98644-1 (April 2008)
Gurtov: *Host Identity Protocol (HIP): Towards the Secure Mobile Internet* 0-470-99790-7 (June 2008)
Boucadair: *Inter-Asterisk Exchange (IAX): Deployment Scenarios in SIP-enabled Networks* 0-470-77072-4 (January 2009)
Fitzek: *Mobile Peer to Peer (P2P): A Tutorial Guide* 0-470-69992-2 (June 2009)
Shelby: *6LoWPAN: The Wireless Embedded Internet* 0-470-74799-4 (November 2009)
Stavdas: *Core and Metro Networks* 0-470-51274-1 (February 2010)

6LoWPAN: The Wireless Embedded Internet

Zach Shelby

Sensinode, Finland

Carsten Bormann

Universität Bremen TZI, Germany

A John Wiley and Sons, Ltd, Publication

This edition first published 2009
© 2009 John Wiley & Sons Ltd

Registered office
John Wiley & Sons Ltd, The Atrium, Southern Gate, Chichester, West Sussex, PO19 8SQ,
United Kingdom.

For details of our global editorial offices, for customer services and for information about how to apply
for permission to reuse the copyright material in this book please see our website at www.wiley.com.

The right of the author to be identified as the author of this work has been asserted in accordance with
the Copyright, Designs and Patents Act 1988.

All rights reserved. No part of this publication may be reproduced, stored in a retrieval system, or
transmitted, in any form or by any means, electronic, mechanical, photocopying, recording or
otherwise, except as permitted by the UK Copyright, Designs and Patents Act 1988, without the prior
permission of the publisher.

Wiley also publishes its books in a variety of electronic formats. Some content that appears in print
may not be available in electronic books.

Designations used by companies to distinguish their products are often claimed as trademarks.
All brand names and product names used in this book are trade names, service marks, trademarks or
registered trademarks of their respective owners. The publisher is not associated with any product or
vendor mentioned in this book. This publication is designed to provide accurate and authoritative
information in regard to the subject matter covered. It is sold on the understanding that the publisher is
not engaged in rendering professional services. If professional advice or other expert assistance is
required, the services of a competent professional should be sought.

Library of Congress Cataloging-in-Publication Data

Shelby, Zach.
    6LoWPAN : the wireless embedded internet / Zach Shelby and Carsten Bormann.
      p. cm.
    Includes bibliographical references and index.
    ISBN 978-0-470-74799-5 (cloth)
1. Wireless Internet. 2. Wireless communication systems–Standards. 3. Low voltage systems. I.
Bormann, Carsten. II. Title.
    TK5103.4885.S52 2009
    621.384–dc22   2009026837

A catalogue record for this book is available from the British Library.

ISBN 9780470747995 (H/B)

Set in 10/12pt Times by Sunrise Setting Ltd, Torquay, UK.
Printed in Great Britain by Antony Rowe, Chippenham, Wiltshire.

Contents

List of Figures   ix

List of Tables   xiii

Foreword   xv

Preface   xvii

Acknowledgments   xix

1 Introduction   1
   1.1 The Wireless Embedded Internet   3
        1.1.1 Why 6LoWPAN?   4
        1.1.2 6LoWPAN history and standardization   6
        1.1.3 Relation of 6LoWPAN to other trends   8
        1.1.4 Applications of 6LoWPAN   9
        1.1.5 Example: facility management 11
   1.2 The 6LoWPAN Architecture   13
   1.3 6LoWPAN Introduction   15
        1.3.1 The protocol stack   16
        1.3.2 Link layers for 6LoWPAN   17
        1.3.3 Addressing 19
        1.3.4 Header format 20
        1.3.5 Bootstrapping   20
        1.3.6 Mesh topologies 22
        1.3.7 Internet integration 23
   1.4 Network Example 24

2 The 6LoWPAN Format   27
   2.1 Functions of an Adaptation Layer 28
   2.2 Assumptions About the Link Layer   29
        2.2.1 Link-layer technologies beyond IEEE 802.15.4 29
        2.2.2 Link-layer service model 30
        2.2.3 Link-layer addressing   31
        2.2.4 Link-layer management and operation   32

2.3   The Basic 6LoWPAN Format   32
2.4   Addressing   34
2.5   Forwarding and Routing   37
      2.5.1   L2 forwarding ("Mesh-Under")   38
      2.5.2   L3 routing ("Route-Over")   40
2.6   Header Compression   41
      2.6.1   Stateless header compression   43
      2.6.2   Context-based header compression   45
2.7   Fragmentation and Reassembly   52
      2.7.1   The fragmentation format   55
      2.7.2   Avoiding the fragmentation performance penalty   59
2.8   Multicast   59

3 Bootstrapping and Security   63
   3.1   Commissioning   64
   3.2   Neighbor Discovery   66
         3.2.1   Forming addresses   67
         3.2.2   Registration   69
         3.2.3   Registration collisions   73
         3.2.4   Multihop registration   77
         3.2.5   Node operation   80
         3.2.6   Router operation   81
         3.2.7   Edge router operation   82
   3.3   Security   83
         3.3.1   Security objectives and threat models   84
         3.3.2   Layer 2 mechanisms   85
         3.3.3   Layer 3 mechanisms   87
         3.3.4   Key management   89

4 Mobility and Routing   91
   4.1   Mobility   92
         4.1.1   Mobility types   92
         4.1.2   Solutions for mobility   94
         4.1.3   Application methods   96
         4.1.4   Mobile IPv6   97
         4.1.5   Proxy Home Agent 100
         4.1.6   Proxy MIPv6   100
         4.1.7   NEMO   102
   4.2   Routing   104
         4.2.1   Overview   104
         4.2.2   The role of Neighbor Discovery   107
         4.2.3   Routing requirements 108
         4.2.4   Route metrics 109
         4.2.5   MANET routing protocols   111
         4.2.6   The ROLL routing protocol   114
         4.2.7   Border routing   119

4.3 IPv4 Interconnectivity 120
    4.3.1 IPv6 transition 121
    4.3.2 IPv6-in-IPv4 tunneling 122

5 Application Protocols   **125**
5.1 Introduction 126
5.2 Design Issues 127
    5.2.1 Link layer 129
    5.2.2 Networking 130
    5.2.3 Host issues 130
    5.2.4 Compression 131
    5.2.5 Security 131
5.3 Protocol Paradigms 132
    5.3.1 End-to-end 132
    5.3.2 Real-time streaming and sessions 132
    5.3.3 Publish/subscribe 133
    5.3.4 Web service paradigms 134
5.4 Common Protocols 134
    5.4.1 Web service protocols 135
    5.4.2 MQ telemetry transport for sensor networks (MQTT-S) 137
    5.4.3 ZigBee compact application protocol (CAP) 139
    5.4.4 Service discovery 141
    5.4.5 Simple network management protocol (SNMP) 142
    5.4.6 Real-time transport and sessions 143
    5.4.7 Industry-specific protocols 144

6 Using 6LoWPAN   **149**
6.1 Chip Solutions 150
    6.1.1 Single-chip solutions 150
    6.1.2 Two-chip solutions 151
    6.1.3 Network processor solutions 151
6.2 Protocol Stacks 152
    6.2.1 Contiki and uIPv6 153
    6.2.2 TinyOS and BLIP 153
    6.2.3 Sensinode NanoStack 154
    6.2.4 Jennic 6LoWPAN 155
    6.2.5 Nivis ISA100 155
6.3 Application Development 156
6.4 Edge Router Integration 159

7 System Examples   **163**
7.1 ISA100 Industrial Automation 164
    7.1.1 Motivation for industrial wireless sensor networks 164
    7.1.2 Complications of the industrial space 165
    7.1.3 The ISA100.11a standard 166
    7.1.4 ISA100.11a data link layer 169

7.2   Wireless RFID Infrastructure 170
        7.2.1   Technical overview 172
        7.2.2   Benefits from 6LoWPAN 173
7.3   Building Energy Savings and Management 174
        7.3.1   Network architecture 174
        7.3.2   Technical overview 174
        7.3.3   Benefits from 6LoWPAN 175

8   Conclusion   177

A   IPv6 Reference   181
A.1   Notation 181
A.2   Addressing 182
A.3   IPv6 Neighbor Discovery 184
A.4   IPv6 Stateless Address Autoconfiguration 188

B   IEEE 802.15.4 Reference   191
B.1   Introduction 191
B.2   Overall Packet Format 192
B.3   MAC-layer Security 194

List of Abbreviations   195

Glossary   203

References   209

Index   219

List of Figures

1.1   Wireless embedded 6LoWPAN device   2
1.2   The Internet of Things vision   4
1.3   The relation of 6LoWPAN to related standards and alliances   7
1.4   Example of a personal fitness monitoring application   10
1.5   Example of an industrial safety application   11
1.6   An example of a facility management system including an automatic
      metering infrastructure (AMI)   12
1.7   The 6LoWPAN architecture   14
1.8   IP and 6LoWPAN protocol stacks   16
1.9   IPv6 edge router with 6LoWPAN support   17
1.10 6LoWPAN header compression example   21
1.11 6LoWPAN/UDP compressed headers (6 bytes)   21
1.12 Standard IPv6/UDP headers (48 bytes)   22
1.13 A 6LoWPAN example   24

2.1   Uncompressed IPv6 packet with 6LoWPAN header   33
2.2   Composition of an EUI-64   35
2.3   Composition of an IPv6 address from an EUI-64: U is the inverted L bit . . .   36
2.4   Interface identifier for 16-bit short addresses   37
2.5   The IP routing model   38
2.6   The LoWPAN routing model (L3 routing, "Route-Over")   38
2.7   DLL mesh forwarding below the LoWPAN adaptation layer   39
2.8   LoWPAN adaptation layer mesh forwarding   39
2.9   Mesh addressing type and header   40
2.10 Hop-by-hop header compression with two different header compression
      methods   42
2.11 HC1-compressed IPv6 packet: without and with HC2   43
2.12 IPv6 header: non-address fields   44
2.13 Best-case HC1-/HC2-compressed IPv6 packet   46
2.14 LOWPAN_IPHC header   47
2.15 LOWPAN_IPHC traffic class and flow label compression   48
2.16 LOWPAN_NHC base header for UDP   50
2.17 LOWPAN_NHC base header for IPv6 extension headers   50
2.18 LOWPAN_NHC port number compression   51
2.19 Best-case LOWPAN_IPHC IPv6 packet   52
2.20 Globally routable best-case LOWPAN_IPHC IPv6 packet   52

2.21 Fragmentation fields in the IPv4 Header   53
2.22 IPv6 fragment header   55
2.23 Non-initial 6LoWPAN fragment   56
2.24 Initial 6LoWPAN fragment   56
2.25 The LOWPAN_BC0 broadcast header   60
2.26 IP multicast address to 16-bit short address mapping   61

3.1   6LoWPAN information option   68
3.2   Router Advertisement dissemination   68
3.3   6LoWPAN summary option   69
3.4   Basic router discovery and registration process with an edge router   70
3.5   Node registration/confirmation message format   70
3.6   Address option format   71
3.7   Example: Node Registration with two address options   73
3.8   Example: Node Confirmation with two address options   74
3.9   Example: the second address option in a refresh NR message   74
3.10 The transaction ID (TID) sequence number lollipop   76
3.11 Router performing ICMP relay on the NR/NC messages   78
3.12 The registration process: multihop operation   78
3.13 Extended LoWPAN operation as a binding moves to a new edge router   83
3.14 Owner interface identifier option   83
3.15 Encapsulating security payload (ESP) packet format   88
3.16 ESP payload encrypted with AES/CCM   89

4.1   An industrial asset management application where mobility is common . . .   92
4.2   The difference between micro-mobility and macro-mobility   94
4.3   Network mobility example   95
4.4   Example of Mobile IPv6 used with 6LoWPAN   99
4.5   Example of a proxy Home Agent located on an edge router 101
4.6   Example of PMIPv6 with 6LoWPAN 102
4.7   Example of the basic NEMO protocol working with 6LoWPAN 103
4.8   Stack view of forwarding inside the LoWPAN and across the edge router . . . 105
4.9   Topology view of forwarding inside the LoWPAN and across the edge router 106
4.10 Example of reactive distance-vector routing 113
4.11 The ROLL architecture 117
4.12 Examples of upstream and downstream forwarding with ROLL 118
4.13 Border routing example 121
4.14 Configured IPv6-in-IPv4 tunneling example 123

5.1   Applications process communication occurs through Internet sockets 126
5.2   The relationship of common IP protocols 128
5.3   Application design issues to consider and where they occur in a LoWPAN . . 129
5.4   End-to-end and proxied application protocol paradigms 133
5.5   Typical structure of web service content over HTTP/TCP 135
5.6   The MQTT-S architecture used over 6LoWPAN 138
5.7   The MQTT-S message structure 138

5.8   The CAP protocol stack 140
5.9   The RTP base header 143

6.1   An example embedded device using a modular two-chip (MSP430+CC2420)
       design 149
6.2   Single-chip solution architecture 150
6.3   Two-chip solution architecture 151
6.4   Network processor solution architecture 152
6.5   The Contiki architecture 154
6.6   The NanoStack architecture 155
6.7   Example use of a socket-like API 157
6.8   Edge router with a 6LoWPAN network interface 160

7.1   The ISA100 network architecture 168
7.2   Forwarding at the link-layer through the ISA100 protocol stack 168
7.3   The Idesco Cardea system architecture 171
7.4   The wireless communications between Cardea components 173
7.5   The typical network architecture of a LessTricity deployment 175
7.6   The Jennic 6LoWPAN stack with the LessTricity application 176

A.1   IPv6 packet header 181
A.2   IPv6 packet header in box notation 182
A.3   IPv6 link-local address 183
A.4   IPv6 global unicast address 183
A.5   IPv6 multicast address 184
A.6   Flag values for IPv6 multicast addresses 184
A.7   General format of ICMPv6 messages 185
A.8   General format of an ICMPv6 message option 185
A.9   IPv6 Router Advertisement message 186
A.10   IPv6 Router Solicitation message 187
A.11   IPv6 ND prefix information option 187

B.1   Overall structure of the IEEE 802.15.4 data packet 193
B.2   The security subheader in an IEEE 802.15.4 data packet 194

List of Tables

2.1 The two most-significant bits in the dispatch byte 33
2.2 Current and proposed dispatch byte allocations 35
2.3 Address ranges for 16-bit short addresses 37
2.4 HC1 SAE and DAE values 44
2.5 HC1 NH values 44
2.6 S/SAM values, and D/DAM values when M = 0 49
2.7 EID values in LOWPAN_NHC base header 49
2.8 D/DAM values for M = 1 61

3.1 Information content of a Node Registration binding 75
3.2 Processing rules for NR messages 77

4.1 Non-exhaustive summary of requirements identified for ROLL 110
4.2 Routing metrics identified for ROLL 111

7.1 Classes of sensor and control applications 166

A.1 IPv6 addresses in hexadecimal notation 183
A.2 Scope values for IPv6 multicast addresses 184

B.1 Frequency ranges and channels for IEEE 802.15.4 192
B.2 Fields in the fixed header/trailer of the IEEE 802.15.4 data packet 193
B.3 Addressing modes in the IEEE 802.15.4 MAC header 193
B.4 Key identifier modes (KIM) in the IEEE 802.15.4 security subheader 194
B.5 Security level (LVL) in the IEEE 802.15.4 security subheader 194

Foreword

You are holding (or perhaps reading online or in an e-book) a remarkable volume. I have been a proponent of IPv6 and an enthusiastic adopter of sensor networks for some time. I am using a commercially available 6LoWPAN system to monitor my home and especially the wine cellar. You may imagine my positive reaction to the book you are reading now. It is stunningly thorough and takes readers meticulously through the design, configuration and operation of IPv6-based, low-power, potentially mobile radio-based networking.

In reading through this book, I was struck also by the thoughtful framing of issues that reach beyond the specifics of 6LoWPAN and go to the heart of many aspects of Internet protocol design. For example, general problems, such as packet fragmentation, are explained in the context of the standard Internet protocols and then, more particularly, in the context of 6LoWPAN. This technique helps to place issues into broader contexts and takes advantage of knowledge that readers may have already of the Internet Architecture.

Sensor network utility seems to me indisputable and consequently, this book has wide-ranging implications for anyone thinking about the proliferation of sensor networks, the need for significant address space to support them, and their integration into the present IPv4 Internet and the future IPv6 Internet. The special requirements imposed by battery-powered operation, radio-based communication and potentially mobile operation motivate the need for books of this caliber. Whoever said "the devil is in the details" might well have had 6LoWPAN in mind!

I found the sections on mobility particularly helpful and the term "micro-mobility" especially illuminating. Mobility in the Internet's design has long been a problem area and I had been puzzled by this since the original Internet included two mobile packet radio networks (in the San Francisco Bay area and Fort Bragg, North Carolina). It is clear that the mobility conferred by these networks was confined to mobility *within* a given packet radio system, in other words, micro-mobility as defined by the authors. That's the easy kind. The hard kind is when the IP address of the mobile node has to change to reflect a new topological access point into the Internet. It is that kind of mobility that has not been well served by present-day Internet protocols. There is still much work to be done to handle this better. The need for a *Home Agent* is a reflection of the awkwardness of IP mobility in general. The 6LoWPAN design does the best it can to deal with this, within the present-day IPv6 architecture.

Routing in low-power, lossy environments has been taken up by the ROLL working group in the Internet Engineering Task Force. In addition, the Mobile Ad-Hoc Network (MANET) working group has also tackled aspects of this problem. These sections of the book are extremely valuable for their pedagogical utility to say nothing of the practical consideration they give to this vexing problem area.

I found the sections on Applications (Chapter 5) especially interesting since that is where all the real action is. Figure 5.3 is a beautiful example of using simple diagrams to localize problem areas and issues. This chapter highlighted for me the importance of matching the applications to the underlying capability of the network(s) through which the application must operate. If end-to-end connectivity is not guaranteed, applications need to incorporate awareness of this if they are to operate successfully and effectively, for example. Blindly layering protocols accustomed to reliable, speedy and sequenced delivery on critical network components that cannot provide such guarantees will generally produce unsatisfactory results.

As we enter into a period where sensors networks become an integral part of energy management, building automation, and other applications, it is highly desirable to standardize application infrastructure to enable interoperability among systems from many vendors. In Chapter 5, we encounter ideas that enable the experience obtained from the proprietary ZigBee space to be adapted to operate in the UDP/IP/6LoWPAN space. It is encouraging to see such efforts at synthesizing commonality to increase interoperability and to enable competitive offerings. The so-called CAP protocol is the key element at work and strikes me as an important contribution to the Internet protocol library. The chapter finishes with a very useful compendium and summary of a variety of proprietary protocols that ultimately will have to be adapted to work in a more standard Internet environment to be broadly useful.

The convergence of ZigBee protocols with Internet-oriented ones, in the 6LoWPAN context, and the creation of the IP for Smart Objects (IPSO) alliance are healthy indications that the ad hoc solutions for low-power networking are beginning to coalesce into interoperable designs that can become the core of the Internet of Things. I cannot see all the ramifications of this emerging consensus but it is fair to say that it will deliver an information-rich environment in which to invent new applications and provide feedback that will enable wiser choices leading to an environmentally smarter society.

Vint Cerf
Vice-president and Chief Internet Evangelist, Google

Preface

The *Internet of Things* is considered to be the next big opportunity, and challenge, for the Internet engineering community, users of technology, companies and society as a whole. It involves connecting embedded devices such as sensors, home appliances, weather stations and even toys to Internet Protocol (IP) based networks. The number of IP-enabled embedded devices is increasing rapidly, and although hard to estimate, will surely outnumber the number of personal computers (PCs) and servers in the future. With the advances made over the past decade in microcontroller, low-power radio, battery and microelectronic technology, the trend in the industry is for smart embedded devices (called smart objects) to become IP-enabled, and an integral part of the latest services on the Internet. These services are no longer cyber, just including data created by humans, but are to become very connected to the physical world around us by including sensor data, the monitoring and control of machines, and other kinds of physical context. We call this latest frontier of the Internet, consisting of wireless low-power embedded devices, the *Wireless Embedded Internet*. Applications that this new frontier of the Internet enable are critical to the sustainability, efficiency and safety of society and include home and building automation, healthcare, energy efficiency, smart grids and environmental monitoring to name just a few.

Standards for the Internet are set by the Internet Engineering Task Force (IETF). A new set of IETF standards for IPv6 over low-power wireless area networks (6LoWPAN) will be a key technology for the Wireless Embedded Internet. Originally WPAN stood for wireless Personal area network, a term inherited from IEEE 802.15.4, which is no longer descriptive for the wide range of applications for 6LoWPAN. In this book we use the term low-power wireless area network (LoWPAN). This book is all about 6LoWPAN, giving a complete overview of the technology, its application, related standards along with real-life deployment and implementation considerations. The low-power networking industry, from ZigBee ad hoc control to industrial automation standards like ISA100, is quickly converging to the use of IP technology, and IPv6 in particular. 6LoWPAN plays an important role in this convergence of heterogeneous technologies, interest groups and applications behind Internet technology.

This book is meant to be an introduction and reference to understanding and applying 6LoWPAN for use by experts in embedded systems, networking or Internet applications, by both undergrad and postgrad engineering students as well as by lecturers. The book has been designed, along with its accompanying material, to be directly used as the basis for an intensive short course on 6LoWPAN, or as a module in a full course.

Please visit the official web-site of the book at http://6lowpan.net. There you will find accompanying material for the book, including course material and 6LoWPAN programming exercises. An interactive 6LoWPAN blog by the authors, along with other 6LoWPAN material is also available at the site. We would love to hear your comments, ideas and advice.

In order to get the most out of this book it is recommended that the reader has background understanding of the Internet architecture [RFC1958, RFC3439], IPv6 [RFC2460, RFC4291, RFC4861] along with wireless communication basics. The book makes wide use of references to Internet Engineering Task Force (IETF) Request For Comments (RFC) [RFCxxxx] and Internet-Draft (I-D) [ID-xx-xx-xx] documents, which are accessible freely and easily at http://www.ietf.org. Keep in mind that Internet-Drafts are a work in progress as part of the IETF standardization process, and change frequently before possibly becoming an RFC.

The book is organized as follows. Chapter 1 gives an overview of the Wireless Embedded Internet, 6LoWPAN and its architecture. Chapter 2 introduces the 6LoWPAN format, features and addressing in detail, and explains how it works in practice. Chapter 3 looks at bootstrapping 6LoWPAN networks using Neighbor Discovery, and security issues related to these networks. Chapter 4 looks at the important topic of mobility issues and routing, both inside 6LoWPAN networks and with the Internet. Application protocols are considered in Chapter 5. Finally implementation issues related to using 6LoWPAN in embedded devices and routers are covered in Chapter 6 and several examples of systems using 6LoWPAN are given in Chapter 7 including the ISA100 standard. Conclusions and future challenges are discussed in Chapter 8. For ease of reference, appendices are included with basic information on IPv6 (Appendix A) and IEEE 802.15.4 (Appendix B).

As telecommunications and Internet engineering is a mine field of special terminology, many terms often conflicting, we have included a glossary of the most important terms for understanding the subject as a reference at the end of the book. The relevant IETF documents also include terminology sections in the beginning which can be useful for understanding.

Finally, we make use of IETF style packet header diagrams, which for historical and practical reasons are (even today) drawn using ASCII art! This makes it much easier for the reader to reference IETF documents for further reading on protocol details. An explanation of this format is included in Appendix A.

Zach Shelby and Carsten Bormann
Sensinode, Finland and Universität Bremen TZI, Germany

Acknowledgments

We would first like to thank the people that have collaborated with us on this book. Special acknowledgment goes to Geoff Mulligan, co-chair of the 6LoWPAN working group at the IETF and chairman of the IPSO Alliance, who has collaborated with us since the beginning of this project. Geoff was key in encouraging and helping develop the concept for the book, provided comments on the manuscript and contributed Section 7.1 (ISA100 Industrial Automation) of the book. We give thanks to Anthony Schoofs, Oliver Laumann, Pascal Thubert, and Vint Cerf, who have given us very helpful comments during the preparation of the manuscript.

The good people at Wiley have been critical for making this a smooth project. We especially thank Tiina Ruonamaa for giving us the courage to turn a rough idea into a book proposal and finally a book.

The unique culture and resourceful constituency of the Internet Engineering Task Force have made this technology possible in the first place, and have given both of us great experiences over the years. We would like to thank the dedicated and highly motivated people in the 6LoWPAN and ROLL working groups, who have spent the past few years, and several GB of email traffic, developing the concepts and protocols for 6LoWPAN. We would especially like to thank Dominique Barthel, Anders Brandt, Ian Chakeres, Samita Chakrabarti, David Culler, Mischa Dohler, Carles Gomez, Jonathan Hui, Dominik Kaspar, Eunsook "Eunah" Kim, Nandakishore Kushalnagar, Philip Levis, Jerry Martocci, Gabriel Montenegro, Bob Moskowitz, Charlie Perkins, Kris Pister, Anthony Schoofs, Pascal Thubert, JP Vasseur, Thomas Watteyne and Tim Winter. Carsten would like to single out the support he got, and the immense technical input, from IETF area directors Thomas Narten, Margaret Wasserman, Mark Townsley, Jari Arkko, and most recently Ralph Droms, from the 6LoWPAN WG advisor Erik Nordmark and 6LoWPAN WG secretary Christian Peter Pii Schumacher, and especially (again) from his esteemed co-chair Geoff Mulligan.

Zach would like to thank his colleagues at Sensinode, the Center for Wireless Communications, the Technical Research Center of Finland and in the SENSEI project. The people active in the IP Smart Object Alliance also deserve thanks; IPSO has been great for the success of embedded IP in the marketplace. He would especially like to thank Prof. Petri Mähonen and Prof. Carlos Pomalaza-Raez for bringing him under their wings and for their endless encouragement and inspiration. He sends warm thanks to his family, especially to his wife Sari for putting up with this project, and two little girls Selna and Alme who already expect embedded gadgets to do anything possible. Special thanks to Marketta, without whose help this book would not have been possible.

Carsten would like to thank his colleagues at Universität Bremen TZI and at the companies NetCS, Tellique and Lysatiq for sharpening his view on what is a viable, implementable and successful protocol and for fostering the space in which this work could prosper. He especially would like to thank Prof. Jörg Ott, with whom he has had many extremely rewarding interactions about protocol design. Infinite thanks to Prof. Ute Bormann, who has helped him on so many levels, both as a colleague at TZI and as a wonderful wife.

1

Introduction

The Internet has been a great success over the past 20 years, growing from a small academic network into a global, ubiquitous network used regularly by over 1.4 billion people. It was the power of the Internet paradigm, tying heterogeneous networks together, and the innovative World Wide Web (WWW) model of uniform resource locators (URLs), the hypertext transfer protocol (HTTP) and universal content markup with the hypertext markup language (HTML) that made this possible. Grass-roots innovation has however been the most powerful driver behind the Internet success story. The Internet is open to innovation like no other telecommunication system before it. This has allowed all groups involved, from Internet architects to communication engineers, IT staff and everyday users to innovate, quickly adding new protocols, services and uses for Internet technology.

As the Internet of routers, servers and personal computers has been maturing, another Internet revolution has been going on – *The Internet of Things*. The vision behind the Internet of Things is that embedded devices, also called *smart objects*, are universally becoming IP enabled, and an integral part of the Internet. Examples of embedded devices and systems using IP today range from mobile phones, personal health devices and home automation, to industrial automation, smart metering and environmental monitoring systems. The scale of the Internet of Things is already estimated to be immense, with the potential of trillions of devices becoming IP-enabled. The impact of the Internet of Things will be significant, with the promise of better environmental monitoring, energy savings, smart grids, more efficient factories, better logistics, better healthcare and smart homes.

The Internet of Things revolution started in the 1990s with industrial automation systems. Early proprietary networks in industrial automation were quickly replaced by different forms of industrial Ethernet, and Internet protocols became widely used between embedded automation devices and back-end systems. This trend has continued in all other automation segments, with Ethernet and IP becoming ubiquitous. Machine-to-machine (M2M) telemetry made a breakthrough already in the early 2000s, with the use of cellular modems and IP to monitor and control a wide range of equipment from vending machines to water pumps. Building automation systems have gone from legacy control to making wide use of wired IP communications through the *Building Automation and Control Network* (BACnet) and

6LoWPAN: The Wireless Embedded Internet   Zach Shelby and Carsten Bormann
© 2009 John Wiley & Sons, Ltd

Open Building Information Exchange (oBIX) standards. More recently, automatic metering infrastructures and smart grids are being deployed at a rapid rate, largely depending on the scalability and universal availability of IP technology. Finally, mobile phones have become almost universally IP-enabled embedded devices currently making up the largest body of devices belonging to the Internet of Things.

An equally important development has been happening in the services that are used to monitor and control embedded devices. Today these services are almost universally built on Internet technology, and more commonly are implemented using web-based services. *Web Service* technologies have completely changed the way business and enterprise applications are designed and deployed. It is this combination of Internet-connected embedded devices and Web-based services which makes the Internet of Things a powerful paradigm.

Hundreds of millions of embedded devices are already IP-enabled, but the Internet of Things is still in its infancy in 2009. Although the capabilities of processor, power and communications technology have continuously increased, so has the complexity of communications standards, protocols and services. Thus, so far, it has been possible to use Internet capabilities in only the most powerful embedded devices. Additionally, low-power wireless communications limits the practical bandwidth and duty-cycle available. Throughout the 1990s and early 2000s we have seen a large array of proprietary low-power embedded wireless radio and networking technologies. This has fragmented the market and slowed down the deployment of such technology.

The Institute of Electrical and Electronics Engineers (IEEE) released the 802.15.4 low-power wireless personal area network (WPAN) standard in 2003, which was a major milestone, providing the first global low-power radio standard. Soon after, the ZigBee Alliance developed a solution for ad hoc control networks over IEEE 802.15.4, and has produced a lot of publicity about the applications of wireless embedded technology. ZigBee and proprietary networking solutions that are vertically bound to a link-layer and application profiles only solve a small portion of the applications for wireless embedded networking. They also have problems with scalability, evolvability and Internet integration. A new paradigm was needed to enable low-power wireless devices with limited processing capabilities (see Figure 1.1) to participate in the Internet of Things, forming what we call the *Wireless Embedded Internet*.

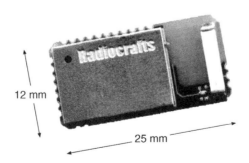

Figure 1.1 Wireless embedded 6LoWPAN device.

This book introduces a set of Internet standards which enable the use of *IPv6 over low-power wireless area networks* (6LoWPAN)[1], which is the key to realizing the Wireless Embedded Internet. 6LoWPAN breaks down the barriers to using IPv6 in low-power, processing-limited embedded devices over low-bandwidth wireless networks. IPv6, which is the newest version of the Internet Protocol, was developed in the late 1990s as a solution to the rapid growth and challenges facing the Internet. The further growth of the Internet of Things will be made possible thanks to IPv6.

In this chapter we give an overview of 6LoWPAN. First the Internet of Things is introduced, followed by the ideas behind 6LoWPAN, IETF standardization, related trends and applications of 6LoWPAN technology in Section 1.1. The overall 6LoWPAN architecture is then introduced in Section 1.2. A comprehensive overview of 6LoWPAN basic mechanisms and the link-layer are given in Section 1.3, followed by a 6LoWPAN network example in Section 1.4.

1.1   The Wireless Embedded Internet

What is the Internet of Things in practice? Maybe the simplest definition is that the Internet of Things encompasses all the embedded devices and networks that are natively IP-enabled and Internet-connected, along with the Internet services monitoring and controlling those devices. Figure 1.2 shows an illustration of the Internet of Things vision.

Today's Internet is made up of a *core Internet* of backbone routers and servers, including millions of nodes (any kind of network device) in total. The core Internet changes rarely and has extremely high capacity. The vast majority of today's Internet nodes are in what is sometimes called the *fringe Internet*. The fringe Internet includes all the personal computers, laptops and local network infrastructure connected to the Internet. This fringe changes rapidly, and is estimated to have up to a billion nodes. In 2008 it was estimated that the Internet had approximately 1.4 billion regular users, and Google announced that over a *trillion* unique URLs existed in their search indexes. The growth of the fringe is dependent on the number of Internet users and the personal devices used by them. The Internet of Things, sometimes referred to as the *embedded fringe*, is the biggest challenge and opportunity for the Internet today. It is made up of the IP-enabled embedded devices connected to the Internet, including sensors, machines, active positioning tags, radio-frequency identification (RFID) readers and building automation equipment to name but a few. The exact size of the Internet of Things is hard to estimate, as its growth is not dependent on human users. It is assumed that the Internet of Things will soon exceed the rest of the Internet in size (number of nodes) and will continue growing at a rapid rate. The long-term potential size of the Internet of Things is in trillions of devices. The greatest growth potential in the future comes from embedded, low-power, wireless devices and networks that until now have not been IP-enabled – the Wireless Embedded Internet. In 2008 the IP Smart Objects (IPSO) Alliance [IPSO] was formed by industry leaders to promote the use of Internet protocols by smart objects and the Internet of Things through marketing, education and interoperability.

The Wireless Embedded Internet is a subset of the Internet of Things, and the main subject of this book. We define the Wireless Embedded Internet to include resource-limited

---

[1] The 6LoWPAN acronym has been redefined on purpose in this book, as "Personal" is no longer relevant to the technology. WPAN originally referred to IEEE 802.15.4 Wireless Personal Area Network.

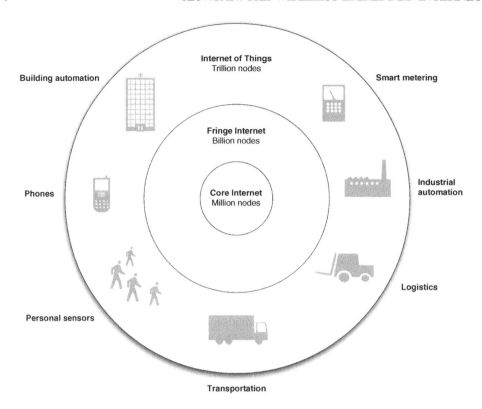

Figure 1.2 The Internet of Things vision.

embedded devices, often battery powered, connected by low-power, low-bandwidth wireless networks to the Internet. 6LoWPAN was developed to enable the Wireless Embedded Internet by simplifying IPv6 functionality, defining very compact header formats and taking the nature of wireless networks into account [6LoWPAN].

1.1.1 Why 6LoWPAN?

There are a huge range of applications which could benefit from a Wireless Embedded Internet approach. Today these applications are implemented using a wide range of proprietary technologies which are difficult to integrate into larger networks and with Internet-based services. The benefits of using Internet protocols in these applications, and thus integrating them with the Internet of Things include [RFC4919]:

- IP-based devices can be connected easily to other IP networks without the need for translation gateways or proxies.

- IP networks allow the use of existing network infrastructure.

- IP-based technologies have existed for decades, are very well known, and have been proven to work and scale. The *socket API* (application programming interface) is one of the most well-known and widely used APIs in the world.

- IP technology is specified in an open and free way, with standards processes and documents available to anyone. The result is that IP technology encourages innovation and is better understood by a wider audience.

- Tools for managing, commissioning and diagnosing IP-based networks already exist (although many management protocols need optimization for direct use with 6LoW-PAN Nodes as we will discuss in Chapter 5).

Until now only powerful embedded devices and networks have been able to participate natively with the Internet. Direct communication with traditional IP networks requires many Internet protocols, often requiring an operating system to deal with the complexity and maintainability. Traditional Internet protocols are demanding for embedded devices for the following reasons:

Security: IPv6 includes optional support for IP Security (IPsec) [RFC4301] authentication and encryption, and web services typically make use of secure sockets or transport layer security mechanisms. These techniques may be too complex, especially for simple embedded devices.

Web services: Internet services today rely on web-services, mainly using the transmission control protocol (TCP), HTTP, SOAP and XML with complex transaction patterns.

Management: Management with the simple network management protocol (SNMP) and web-services is often inefficient and complex.

Frame size: Current Internet protocols require links with sufficient frame length (minimum of 1280 bytes for IPv6), and heavy application protocols require substantial bandwidth.

These requirements have in practice limited the Internet of Things to devices with a powerful processor, an operating system with a full TCP/IP stack, and an IP-capable communication link. Typical embedded Internet devices today include industrial devices with Ethernet interfaces, M2M gateways with cellular modems, and advanced smart phones. A large majority of embedded applications involve limited devices, with low-power wireless and wired network communications. Wireless embedded devices and networks are particularly challenging for Internet protocols:

Power and duty-cycle: Battery-powered wireless devices need to keep low *duty cycles* (the percentage of time active). The basic assumption of IP is that a device is always connected.

Multicast: Wireless embedded radio technologies, such as IEEE 802.15.4, do not typically support multicast, and flooding in such a network is wasteful of power and bandwidth. Multicast is crucial to the operation of many IPv6 features.

Mesh topologies: The applications of wireless embedded radio technology typically benefit from multihop mesh networking to achieve the required coverage and cost efficiency. Current IP routing solutions may not easily be applicable to such networks (discussed at length in Chapter 4).

Bandwidth and frame size: Low-power wireless embedded radio technology usually has limited bandwidth (on the order of 20–250 kbit/s) and frame size (on the order of 40–200 bytes). In mesh topologies, bandwidth further decreases as the channel is shared and is quickly reduced by multihop forwarding. The IEEE 802.15.4 standard has a 127-byte frame size, with layer-2 payload sizes as low as 72 bytes. The minimum frame size for standard IPv6 is 1280 bytes [RFC2460], thus requiring fragmentation.

Reliability: Standard Internet protocols are not optimized for low-power wireless networks. For example, TCP is not able to distinguish between packets dropped because of congestion or packets lost on wireless links. Further unreliability occurs in wireless embedded networks because of node failure, energy exhaustion and sleep duty cycles.

The IETF 6LoWPAN working group [6LoWPAN] was created to tackle these problems, and to specifically enable *IPv6* to be used with wireless embedded devices and networks. Features of the IPv6 design such as a simple header structure, and its hierarchical addressing model, made it ideal for use in wireless embedded networks with 6LoWPAN. Additionally, by creating a dedicated group of standards for these networks, the minimum requirements for implementing a lightweight IPv6 stack with 6LoWPAN could be aligned with the most minimal devices. Finally by designing a version of Neighbor Discovery (ND) specifically for 6LoWPAN, the particular characteristics of low-power wireless mesh networks could be taken into account. The result of 6LoWPAN is the efficient extension of IPv6 into the wireless embedded domain, thus enabling *end-to-end* IP networking and features for a wide range of embedded applications. Refer to [RFC4919] for the detailed assumptions, problem statement and goals of early 6LoWPAN standardization. Although 6LoWPAN was targeted originally at IEEE 802.15.4 radio standards and assumed layer-2 mesh forwarding [RFC4944], it was later generalized for all similar link technologies, with additional support for IP routing in [ID-6lowpan-hc, ID-6lowpan-nd].

1.1.2   6LoWPAN history and standardization

6LoWPAN is a set of standards defined by the Internet Engineering Task Force (IETF), which creates and maintains all core Internet standards and architecture work. A straightforward technical definition of 6LoWPAN would be:

> 6LoWPAN standards enable the efficient use of IPv6 over low-power, low-rate wireless networks on simple embedded devices through an adaptation layer and the optimization of related protocols.

The IETF 6LoWPAN working group was officially started in 2005, although the history of embedded IP goes back farther. Throughout the 1990s it was assumed that Moore's law would advance computing and communication capabilities so rapidly that soon any embedded device could implement IP protocols. Although partially true, and the Internet of Things has grown rapidly, it did not hold for cheap, low-power microcontrollers and low-power wireless radio technologies. The vast majority of simple embedded devices still make use of 8-bit and 16-bit microcontrollers with very limited memory, as they are low-power, small and cheap. At the same time, the physical trade-offs of wireless technology have resulted in short-range, low-power wireless radios which have limited data rates, frame sizes and duty cycles such as in the IEEE 802.15.4 standard. Early work on minimizing Internet protocols for use with low-power microcontrollers and wireless technologies includes μIP from the Swedish Institute of Computer Science [Dunkels03] and NanoIP from the Centre for Wireless Communications [Shel03]. The IEEE 802.15.4 standard released in 2003 was the biggest factor leading to 6LoWPAN standardization. For the first time a global, widely supported standard for low-power wireless embedded communications was available [IEEE802.15.4]. The popularity of this new standard gave the Internet community the needed encouragement to standardize an IP adaptation for such wireless embedded links.

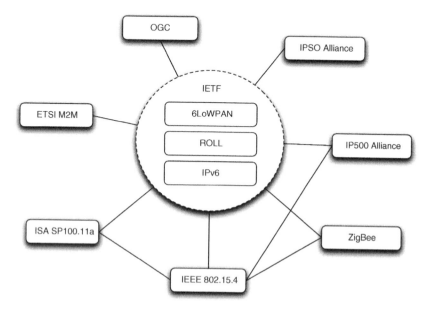

Figure 1.3 The relation of 6LoWPAN to related standards and alliances.

The first 6LoWPAN specifications were released in 2007, first with an informational RFC [RFC4919] specifying the underlying requirements and goals of the initial standardization, and then with a standard track RFC [RFC4944] specifying the 6LoWPAN format and functionality. Through experience with implementations and deployments, the 6LoWPAN working group continued with improvements to header compression [ID-6lowpan-hc], 6LoWPAN Neighbor Discovery [ID-6lowpan-nd], use cases [ID-6lowpan-uc] and routing requirements [ID-6lowpan-rr]. In 2008 a new IETF working group was formed, *Routing over Low-power and Lossy Networks* (ROLL)[ROLL]. This working group specifies routing requirements and solutions for low-power, wireless, unreliable networks. Although not restricted to use with 6LoWPAN, that is one main target.

In 2008 ISA began standardization of a wireless industrial automation system called SP100.11a (also known as ISA100), which is based on 6LoWPAN. An overview of ISA100 is given in Chapter 7. Recent activities related to 6LoWPAN include the IP for Smart Objects (IPSO) Alliance founded in 2008 to promote the use of IP in smart objects and Internet of Things business [IPSO], and the IP500 Alliance which is developing a recommendation for 6LoWPAN over IEEE 802.15.4 sub-GHz radio communications [IP500]. Figure 1.3 shows the relations between related standards bodies and alliances. The Open Geospatial Consortium (OGC) specifies IP-based solutions for geospatial and sensing applications. In 2009 the European Telecommunication Standards Institute (ETSI) [ETSI] started a working group for standardizing M2M, which includes an end-to-end IP architecture compatible with 6LoWPAN.

1.1.3   Relation of 6LoWPAN to other trends

There are several other trends to take into consideration when thinking about the Internet of Things. These include ZigBee, machine-to-machine (M2M) communications, the Future Internet, and wireless sensor networks (WSNs). This section looks into how each of these trends relates to the Internet of Things and 6LoWPAN in particular.

ZigBee is a protocol specification from an industry special interest group called the ZigBee Alliance, specializing in ad hoc control [ZigBee]. ZigBee was started in 2003 in conjunction with IEEE 802.15.4 standardization [IEEE802.15.4], and specifies a vertical protocol stack solution with similarities to Bluetooth. The protocol mainly makes use of IEEE 802.15.4 features, adding ad hoc networking, service discovery and application protocol profiles on top of that. ZigBee has been successful for multi-vendor ad hoc applications such as home automation. The ZigBee approach does have several down-sides, including reliance on a single wireless link technology, tight coupling with application profiles, along with Internet integration and scalability limitations. In 2009 the ZigBee Alliance announced that ZigBee will start to integrate IETF standards such as 6LoWPAN and ROLL into its future specifications. Earlier work has shown how ZigBee application profiles can be carried over UDP/IP and 6LoWPAN [ID-tolle-cap], which is covered in more detail in Section 5.4.3. The integration of IP technology into ZigBee provides a much wider range of networking possibilities, beyond just ad hoc control.

Machine-to-machine (M2M) communications has become a popular industry term for the remote monitoring and control of machines over the Internet. Traditionally, M2M systems include M2M modules (usually a cellular modem) integrated into embedded devices together with an Internet-based back-end system. The M2M module measures and controls the device, and communicates over IP with the back-end M2M service. More recently, M2M gateways to local embedded networked devices have become more common. Thanks to native IP, 6LoWPAN networks can be connected to M2M services through simple routers and thus 6LoWPAN can be considered to be a natural extension of M2M. Machine-to-machine communications has been an important driving force in the development and growth of the Internet of Things, which has continued with the ETSI M2M standardization effort.

The Future Internet [Bauge08] is a term used to describe research into what the Internet architecture and protocols could look like in 10–20 years. The US National Science Foundation has a long-term initiative on Future Internet Design (FIND) which covers network architecture, principles as well as mechanism design [FIND]. Several European projects specialize in Future Internet research, for example the EU 4WARD project [4WARD], in cooperation with the European Future Internet Assembly [FIAssembly]. Although most of the research related to Future Internet does not consider embedded devices and networks, this aspect is starting to gain interest. The EU SENSEI project [SENSEI] for example specializes in making wireless sensor and embedded networks a part of the global Internet, both current and future. One of the subjects of the project is how wireless embedded networks and 6LoWPAN type functionality can be made an integral part of the Future Internet. Several examples throughout this book are taken from the SENSEI project as it has been doing leading work in this area.

The term Wireless Sensor Network (WSN) comes from an academic movement starting in the mid 1990s into research on low-power ad hoc wireless networked sensors and actuators. The US government was very interested in the application of low-power sensing in military and security applications, and provided extensive funding for the subject. The research area later developed into a widely popular subject with a large range of applications, and a huge collection of results and trials. These networks have traditionally been thought to be completely isolated, and thus typically Internet compatibility or standards were not taken into consideration. Instead each project has tended to produce its own optimized wireless, network and algorithm solutions. Additionally most of the envisioned applications of sensor networks were created by university researchers, and they most often did not have a real market need. More recently the importance of standards, marketable applications and the importance of Internet services have encouraged the WSN community to become involved with 6LoWPAN standardization and the IPSO Alliance. The result is that a lot of the innovation produced through WSN research is starting to be applied to Wireless Embedded Internet technology, a good example being the IETF ROLL working group.

There is a strong trend of convergence in standardization, industry and research, as indicated above. This convergence is clearly steering towards an Internet-based approach as the requirements of modern-day embedded applications clearly demand it. 6LoWPAN has been a result of and catalyst for convergence to the Internet of Things.

1.1.4 Applications of 6LoWPAN

The reason why there are such a large number of technical solutions in the wireless embedded networking market is that the requirements, scale and market of embedded applications vary wildly. Applications can range from personal health sensor monitoring to large scale facility monitoring, which differ greatly. This is in contrast to PC information technology, which is fairly homogeneous and mainly aimed at home and office environments. The ideal use of 6LoWPAN is in applications where:

- embedded devices need to communicate with Internet-based services,

- low-power heterogeneous networks need to be tied together,

- the network needs to be open, reusable and evolvable for new uses and services, and

- scalability is needed across large network infrastructures with mobility.

Connecting the Internet to the physical world enables a wide range of interesting applications where 6LoWPAN technology may be applicable, for example:

- home and building automation

- healthcare automation and logistics

- personal health and fitness (see Figure 1.4)

- improved energy efficiency

- industrial automation (see Figure 1.5)

- smart metering and smart grid infrastructures

- real-time environmental monitoring and forecasting

- better security systems and less harmful defense systems

- more flexible RFID infrastructures and uses

- asset management and logistics

- vehicular automation

Figure 1.4 Example of a personal fitness monitoring application. (Reproduced by Permission of © SENSEI Consortium.)

One interesting example application of 6LoWPAN is in *facility management*, which is the management of large facilities using a combination of building automation, asset management and other embedded systems. This quickly growing field can benefit from 6LoWPAN, is feasible with today's technology, and has real business demand. For these reasons it is an ideal example, which is introduced in the next section.

Figure 1.5 Example of an industrial safety application. (Reproduced by Permission of © SENSEI Consortium.)

1.1.5 Example: facility management

Facility management is a very interesting application for the Internet of Things, and is one use case that has been examined in detail by the SENSEI project [SENSEI]. It involves the integrated management of building facilities. Facility management services are becoming more common, and are typically web-based. Figure 1.6 shows a facility management use case from the SENSEI project. Wireless embedded networking has a large range of applications in facility management including:

Door access control: Access control involves the use of RFID or active tag based identifiers to control and log the access to different parts of a building automatically.

Building automation: Building automation involves the use of sensors and control to improve the operations and efficiency of a building.

Tracking: Tracking involves the use of active tags on people, equipment and supplies which are tracked by the wireless infrastructure throughout a facility. Tracking results are used in asset management, security and logistics optimization.

Energy reduction: Energy reduction in facilities can be achieved through intelligent lighting control, heating control, ventilation and air conditioning control, and the automatic power control of electric equipment.

Maintenance: The maintainability of facilities can be improved through the remote monitoring of the building itself and the systems in the building which today are typically monitored manually.

Smart metering: The use of resources in large facilities can be reduced and better controlled through more intelligent metering of electricity, gas and water using an *automatic metering infrastructure* (AMI).

The stakeholders in facility management include the providers of intelligent facility management systems and services, users of these services and third parties. The providers of facility management services play an important role as a huge amount of data needs to be collected, processed and leveraged to provide the services required in a beneficial way. The automation systems in facilities may include access control, building automation, tracking, maintenance monitoring and metering systems. Users of facility management include building owners or renters, building users and facility managers. Additionally many third parties are involved with facility management such as security companies, insurance companies and utilities. Some of these stakeholders are identified in Figure 1.6.

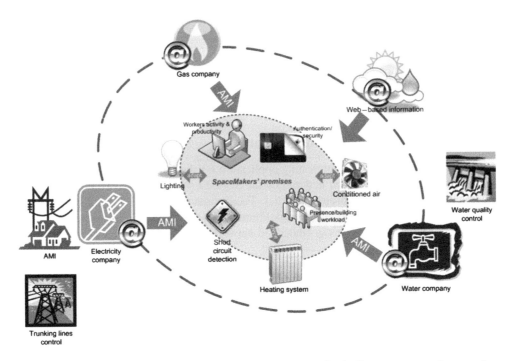

Figure 1.6 An example of a facility management system including an automatic metering infrastructure (AMI). (Reproduced by Permission of © SENSEI Consortium.)

The main rationales for facility management are improvements in energy and resource efficiency, an increase in worker productivity, and more secure and comfortable buildings. Buildings are major consumers of energy: it is estimated that in the EU and the USA, 40 percent of all energy is consumed in the building sector [Baden06, DoE06], and that carbon emissions could be reduced by 22 percent through improved efficiency [2002/91/EC]. For the enterprise users of buildings, an even more important benefit is improved worker efficiency along with better comfort and security in general. Substantial cost savings may be possible through productivity improvements.

Facility management provides many technical challenges for embedded devices and networking. The large range of systems to be integrated needs interoperability between systems, as well as network integration of heterogeneous technology. Furthermore new devices and applications will be added over time, so evolvability is important. The scalability of wireless embedded networking in large buildings is demanding. The density of devices in a single space can reach hundreds of nodes, and there is a mix of fixed and mobile devices across a large area. Battery-powered wireless devices require intelligent networking designed to maximize the lifetime of devices, and thus reduce maintenance. Facility management systems and devices must be cost-efficient, and installation straightforward compared to the long-term benefits achieved through these services. Finally, although privacy is easier in enterprise networks, security is a challenging aspect when applying wireless embedded networking. We will consider how to apply 6LoWPAN to solve networking requirements such as these throughout this book.

1.2   The 6LoWPAN Architecture

The Wireless Embedded Internet is created by connecting islands of wireless embedded devices, each island being a *stub network* on the Internet. A stub network is a network which IP packets are sent from or destined to, but which doesn't act as a transit to other networks. The 6LoWPAN architecture is made up of *low-power wireless area networks* (LoWPANs)[2], which are IPv6 stub networks. The overall 6LoWPAN architecture is presented in Figure 1.7. Three different kinds of LoWPANs have been defined: Simple LoWPANs, Extended LoWPANs, and Ad hoc LoWPANs. A LoWPAN is the collection of 6LoWPAN Nodes which share a common IPv6 address *prefix* (the first 64 bits of an IPv6 address), meaning that regardless of where a node is in a LoWPAN its IPv6 address remains the same. An *Ad hoc LoWPAN* is not connected to the Internet, but instead operates without an infrastructure. A *Simple LoWPAN* is connected through one *LoWPAN Edge Router* to another IP network. A *backhaul link* (point-to-point, e.g. GPRS) is shown in the figure, but this could also be a *backbone link* (shared). An *Extended LoWPAN* encompasses the LoWPANs of multiple edge routers along with a backbone link (e.g. Ethernet) interconnecting them.

LoWPANs are connected to other IP networks through *edge routers*, as seen in Figure 1.7. The edge router plays an important role as it routes traffic in and out of the LoWPAN, while handling 6LoWPAN compression and Neighbor Discovery for the LoWPAN. If the LoWPAN is to be connected to an IPv4 network, the edge router will also handle IPv4 interconnectivity (discussed further in Section 4.3). Edge routers typically have management features tied into overall IT management solutions. Multiple edge routers can be supported in the same LoWPAN if they share a common backbone link.

A LoWPAN consists of nodes, which may play the role of host or router, along with one or more edge routers. The network interfaces of the nodes in a LoWPAN share the same IPv6 prefix which is distributed by the edge router and routers throughout the LoWPAN. In order to facilitate efficient network operation, nodes register with an edge router. These operations are part of *Neighbor Discovery* (ND), which is an important basic mechanism

---

[2]The terms 6LoWPAN and LoWPAN are often used interchangeably. In this book we use 6LoWPAN as a general term for the technology or set of standards, and the term LoWPAN as it is used in the IETF standards: to refer to a specific type of LoWPAN or node.

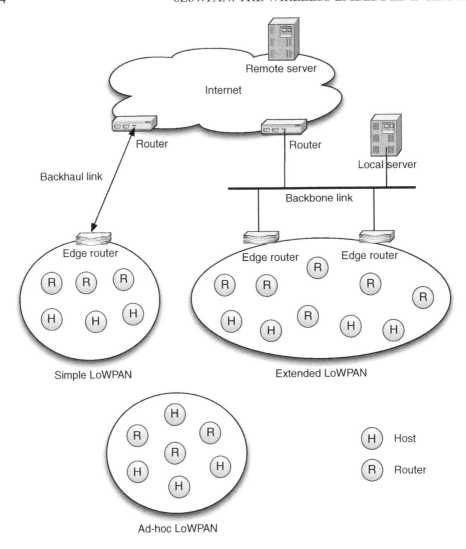

Figure 1.7 The 6LoWPAN architecture.

of IPv6. Neighbor Discovery defines how hosts and routers interact with each other on the same link. LoWPAN Nodes may participate in more than one LoWPAN at the same time (called *multi-homing*), and fault tolerance can be achieved between edge routers. LoWPAN Nodes are free to move throughout the LoWPAN, between edge routers, and even between LoWPANs. Topology change may also be caused by wireless channel conditions, without physical movement. A multihop mesh topology within the LoWPAN is achieved either through link-layer forwarding (called *Mesh-Under*) or using IP routing (called *Route-Over*). Both techniques are supported by 6LoWPAN.

Communication between LoWPAN Nodes and IP nodes in other networks happens in an end-to-end manner, just as between any normal IP nodes. Each LoWPAN Node is identified by a unique IPv6 address, and is capable of sending and receiving IPv6 packets. Typically LoWPAN Nodes support ICMPv6 traffic such as "ping", and use the user datagram protocol (UDP) as a transport. In Figure 1.7 the Simple LoWPAN and Extended LoWPAN Nodes can communicate with either of the servers through their edge router. As the payload and processing capabilities of LoWPAN Nodes are extremely limited, application protocols are usually designed using a simple binary format in a UDP payload. Application protocols suitable for 6LoWPAN are discussed in Chapter 5.

The main difference between a Simple LoWPAN and an Extended LoWPAN is the existence of multiple edge routers in the LoWPAN, which share the same IPv6 prefix and a common backbone link. Multiple LoWPANs can overlap each other (even on the same channel). When moving from one LoWPAN to another, a node's IPv6 address will change. A LoWPAN Edge Router is typically connected to the Internet over a backhaul link such as cellular or DSL [ID-6lowpan-nd]. A network deployment may also choose to use multiple Simple LoWPANs rather than an Extended LoWPAN on a shared backbone link, e.g. for management reasons. This is not a problem if there is low mobility between LoWPANs in the network, or the application does not assume stable IPv6 addresses for nodes. A deployment example of a Simple LoWPAN connected by a backhaul link to the Internet is given in Section 1.4.

In an Extended LoWPAN configuration, as shown on the right-hand side of Figure 1.7, multiple edge routers share a common backbone link and collaborate by sharing the same IPv6 prefix, offloading most Neighbor Discovery messaging to the backbone link [ID-6lowpan-nd]. This greatly simplifies LoWPAN Node operation as IPv6 addresses are stable throughout the Extended LoWPAN and movement between edge routers is very simple. Edge routers also handle IPv6 forwarding on behalf of the nodes. To IP nodes outside the LoWPAN, the LoWPAN Nodes are always reachable regardless of their attachment point in the Extended LoWPAN. This enables large enterprise 6LoWPAN infrastructures to be built, functioning similar to a WLAN (WiFi) access point infrastructure (but at layer 3 instead of layer 2).

6LoWPAN does not require an infrastructure to operate, but may also operate as an Ad hoc LoWPAN [ID-6lowpan-nd]. In this topology, one router must be configured to act as a simplified edge router, implementing two basic functionalities: unique local unicast address (ULA) generation [RFC4193] and handling 6LoWPAN Neighbor Discovery registration functionality. From the LoWPAN Node point of view the network operates just like a Simple LoWPAN, except the prefix advertised is an IPv6 local prefix rather than a global one, and there are no routes outside the LoWPAN.

LoWPAN types and 6LoWPAN Neighbor Discovery operation are covered in detail in Chapter 3. Also refer to the 6LoWPAN Neighbor Discovery document in [ID-6lowpan-nd] for the complete specification.

1.3   6LoWPAN Introduction

This section gives a short but comprehensive introduction to the core 6LoWPAN subjects covered in this book. The protocol stack, link-layer technology, addressing and header format are first explained, followed by bootstrapping, mesh topologies, and Internet integration.

1.3.1   The protocol stack

Figure 1.8 shows the IPv6 protocol stack with 6LoWPAN in comparison with a typical IP protocol stack and the corresponding five layers of the *Internet Model* (the four-layer model of [RFC1122] with a physical layer separated out of the link layer). The Internet Model is sometimes referred to as a "narrow waist" model, as the Internet Protocol ties together a wide variety of *link-layer* technologies with multiple transport and application protocols. A simple IPv6 protocol stack with 6LoWPAN (also called a 6LoWPAN protocol stack) is almost identical to a normal IP stack with the following differences. First of all 6LoWPAN only supports IPv6, for which a small adaptation layer (called the *LoWPAN adaptation layer*) has been defined to optimize IPv6 over IEEE 802.15.4 and similar link layers in [RFC4944]. In practice, 6LoWPAN stack implementations in embedded devices often implement the LoWPAN adaptation layer together with IPv6, thus they can alternatively be shown together as part of the network layer (see Section 6.2 for more about stack implementation issues). The most common transport protocol used with 6LoWPAN is the user datagram protocol (UDP) [RFC0768], which can also be compressed using the LoWPAN format. The transmission control protocol (TCP) is not commonly used with 6LoWPAN for performance, efficiency and complexity reasons. The Internet control message protocol v6 (ICMPv6) [RFC4443] is used for control messaging, for example ICMP echo, ICMP destination unreachable and Neighbor Discovery messages. Application protocols are often application specific and in binary format, although more standard application protocols are becoming available. Application protocols are discussed in detail in Chapter 5.

Adaptation between full IPv6 and the LoWPAN format (described later in this section) is performed by routers at the edge of 6LoWPAN islands, referred to as edge routers. This transformation is transparent, efficient and stateless in both directions. LoWPAN adaptation in an edge router typically is performed as part of the 6LoWPAN network interface driver and is usually transparent to the IPv6 protocol stack itself. Figure 1.9 illustrates one realization of an edge router with 6LoWPAN support. See Section 6.4 for edge router implementation considerations. Inside the LoWPAN, hosts and routers do not actually need to work with full IPv6 or UDP header formats at any point as all compressed fields are implicitly known by each node.

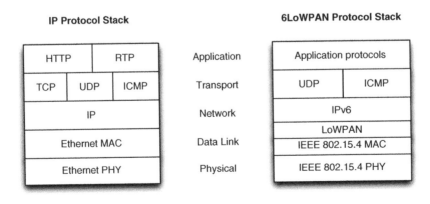

Figure 1.8 IP and 6LoWPAN protocol stacks.

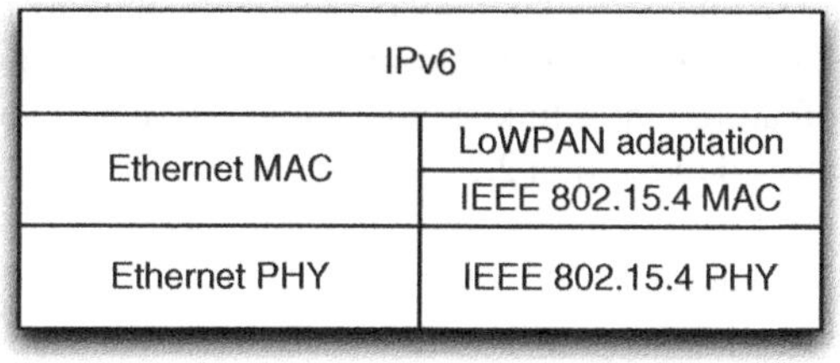

IPv6	
Ethernet MAC	LoWPAN adaptation
	IEEE 802.15.4 MAC
Ethernet PHY	IEEE 802.15.4 PHY

Figure 1.9 IPv6 edge router with 6LoWPAN support.

1.3.2 Link layers for 6LoWPAN

One of the most important functions of the Internet Protocol is the interconnection of heterogeneous links into a single interoperable network, providing a universal "narrow waist". This is equally true for 6LoWPAN and embedded networks, where there are many wireless (and also wired) link-layer technologies in use. The specialized applications of embedded networks require a wider range of communication solutions than typical personal computer networks, which almost universally use Ethernet and WiFi. Luckily the IEEE 802.15.4 standard is the most common 2.4 GHz wireless technology for embedded networking applications, and has been used as a baseline for 6LoWPAN development. Other technologies used with 6LoWPAN include sub-GHz radios, long-range telemetry links and even power-line communications. The requirements and interactions of 6LoWPAN with the link layer are discussed next, along with an introduction to IEEE 802.15.4, a sub-GHz radio and power-line communications.

There is a set of required or recommended features that a link should provide in order to work with Internet protocols. These include framing, addressing, error checking, length indication, some reliability, broadcast and a reasonable frame size. The issues involved with designing a subnetwork for use with IP are discussed in [RFC3819]. 6LoWPAN is designed to be used with a special type of link, and has its own set of link requirements and recommendations.

The most basic requirements for a link layer to support 6LoWPAN are framing, unicast transmission and addressing. Addressing is required to differentiate between nodes on a link, and to form IPv6 addresses which are then elided by 6LoWPAN compression. It is highly recommended that a link supports unique addresses by default (e.g. a 64-bit *extended unique identifier* [EUI-64]), to allow for stateless autoconfiguration. Multi-access links should provide a broadcast service. Multicast service is required by standard IPv6, but not by 6LoWPAN (broadcast is sufficient). IPv6 requires a *maximum transmission unit* (MTU) of 1280 bytes from a link, which 6LoWPAN fulfills by supporting fragmentation at the LoWPAN adaptation layer. A link should provide payload sizes at least 30 bytes in length to be useful (and preferably larger than 60 bytes). Although UDP and ICMP include a simple 16-bit checksum, it is recommended that the link layer also provides strong error checking. Finally, as IPsec may not always be practical for 6LoWPAN, it is highly recommended that links include strong encryption and authentication. The 2006 version of the IEEE 802.15.4

standard actually does not include a "next protocol identifier", making the detection of which protocol is being carried difficult. Although partially dealt with in the LoWPAN format using a *dispatch value*, it is a feature that a link should preferably have. Subnetwork design and link-layer issues are discussed in Section 2.2.

The next sections introduce three link-layer technologies used with 6LoWPAN: IEEE 802.15.4, a sub-GHz ISM band radio and low-rate power line communications.

IEEE 802.15.4

The IEEE 802.15.4 standard [IEEE802.15.4] defines low-power wireless embedded radio communications at 2.4 GHz, 915 MHz and 868 MHz. The first version of the standard was released in 2003, and was then revised in 2006. More recently the IEEE 802.15.4a standard was released, extending 802.15.4 with two new physical layer options: *Chirp Spread Spectrum* at 2.4 GHz and *Ultra Wide-Band* at 3.1–10.6 GHz. Work continues on new features such as MAC improvements in IEEE 802.15.4 Task Group 4e (TG4e), active RFID (TG4f), larger networks (TG4a) and specialized PHYs for China (TG4c) and Japan (TG4d). More information is available on these efforts from [IEEE]. In practice IEEE 802.15.4 at 2.4 GHz is used almost exclusively today as it provides reasonable data rates, and can be used globally. The sub-GHz channels are limited geographically with 915 MHz mainly available in North America and 868 MHz in the European Union (EU). That, combined with the limited data rates and channel selection of sub-GHz IEEE 802.15.4, means that there are only a few chips on the market today. Often more flexible sub-GHz chips tend to be used, as explained in the next section. This trend may yet change, with new sub-GHz applications becoming widespread and efforts like the IP500 Alliance, together with improvements in the latest IEEE 802.15.4 standard for sub-GHz channels.

The 802.15.4 standard provides 20–250 kbit/s data rates depending on the frequency. Channel sharing is achieved using carrier sense multiple access (CSMA), and acknowledgments are provided for reliability. Link-layer security is provided with 128-bit AES encryption. Addressing modes for 64-bit (long) and 16-bit (short) addresses are provided with unicast and broadcast capabilities. The physical layer payload is up to 127 bytes, with 72–116 bytes of payload available after link-layer framing, addressing, and optional security. The MAC can be run in two modes: beaconless mode and beacon-enabled mode. Beaconless mode uses pure CSMA channel access and operates quite like IEEE 802.11 without channel reservations. Beacon-enabled mode uses a hybrid time division multiple access (TDMA) approach, with the possibility of reserving time-slots for critical data. IEEE 802.15.4 includes many mechanisms for forming networks, and for controlling the superframe settings. An IEEE 802.15.4 reference is provided in Appendix B.

Early 6LoWPAN standardization work was originally aimed at the IEEE 802.15.4 standard [RFC4919, RFC4944] and thus assumed that some 802.15.4-specific features such as beacon-enabled mode and association mechanisms would be used along with 802.15.4 device roles. Based on practical experience with [RFC4944] and industry needs, recent 6LoWPAN standardization has been generalized to work with a larger range of link layers and avoids the assumption of IEEE 802.15.4-specific features. The use of 6LoWPAN with IEEE 802.15.4 is covered in more detail in Section 2.2.

Sub-GHz ISM band radios

Sub-GHz radio technologies using the industrial, scientific and medical (ISM) bands for unlicensed operation are especially popular in low-power wireless embedded applications such as telemetry, metering and remote control. The sub-GHz ISM bands cover 433 MHz, 868 MHz and 915 MHz. The main reasons for sub-GHz popularity are the better penetration of lower frequency, resulting in better range compared to 2.4 GHz, and the 2.4 GHz ISM band becoming very crowded in urban environments. One example of a popular sub-GHz chip is the Texas Instruments CC1101 transceiver [CC1101]. This transceiver acts as a reconfigurable radio and is capable of 300–928 MHz operation, with a wide variety of modulations, channel and data rates up to 500 kbit/s. Such a chip can also be used with an external power amplifier for increasing range. The features of the chip include carrier sensing, received signal strength indicator (RSSI) support, and frame sizes up to 250 bytes. The system-on-a-chip version, the CC1110, additionally includes a 128-bit AES encryption hardware engine. This kind of transceiver only provides the physical layer, so the data-link layer is implementation specific and needs to provide e.g. framing, addressing, error checking, acknowledgments and frame length. When designing a link layer for this type of transceiver, the IEEE 802.15.4 frame structure and beaconless mode operation is typically used as a starting point.

Power line communications

6LoWPAN also has interesting uses over special wired communication links, such as low-rate *power line communications* (PLC). Applications of this technology include home automation, energy efficiency monitoring and smart metering. One such system from Watteco [Watteco] uses what is called a watt pulse communication (WPC) technique, greatly reducing the complexity of communications. The data rate of the physical layer provided using WPC is 9.6 kbit/s, and the resulting channel over the power system of a house, building or urban area is multi-access and similar to a wireless CSMA channel. Watteco provides a version of WPC with an emulation of the IEEE 802.15.4 data link layer. This allows 6LoWPAN to be used with PLC in a very similar way to IEEE 802.15.4 and other ISM band radios. With PLC, multihop routing is not an issue as typically all nodes are on the same stable link. Multihop forwarding may be useful to interconnect several PLC subnets, or to integrate PLC and wireless 6LoWPAN islands.

1.3.3 Addressing

IP addressing with 6LoWPAN works just like in any IPv6 network, and is similar to addressing over Ethernet networks as defined by [RFC2464]. IPv6 addresses are typically formed automatically from the prefix of the LoWPAN and the link-layer address of the wireless interfaces. The difference in a LoWPAN is with the way low-power wireless technologies support link-layer addressing; a direct mapping between the link-layer address and the IPv6 address is used for achieving compression. This will be explained in Section 1.3.4.

Low-power wireless radio links typically make use of flat link-layer addressing for all devices, and support both unique long addresses (e.g. EUI-64) and configurable short addresses (usually 8–16 bits in length). The IEEE 802.15.4 standard, for example, supports unique EUI-64 addresses carried in all radio chips, along with configurable 16-bit short addresses. These networks by nature also support broadcast (address 0xFFFF in IEEE 802.15.4), but do not support native multicast.

IPv6 addresses are 128 bits in length, and (in the cases relevant here) consist of a 64-bit prefix part and a 64-bit *interface identifier* (IID) [RFC4291]. *Stateless address autoconfiguration* (SAA) [RFC4862] is used to form the IPv6 interface identifier from the link-layer address of the wireless interface as per [RFC4944]. For simplicity and compression, 6LoWPAN networks assume that the IID has a direct mapping to the link-layer address, therefore avoiding the need for address resolution. The IPv6 prefix is acquired through Neighbor Discovery Router Advertisement (RA) messages [ID-6lowpan-nd] as on a normal IPv6 link. The construction of IPv6 addresses in 6LoWPAN from known prefix information and known link-layer addresses, is what allows a high header compression ratio. 6LoWPAN addressing is discussed in detail in Chapter 2. A reference for IPv6, including the IPv6 addressing model, is provided in Appendix A.

1.3.4   Header format

The main functionality of 6LoWPAN is in its LoWPAN adaptation layer, which allows for the compression of IPv6 and following headers such as UDP along with fragmentation and mesh addressing features. 6LoWPAN headers are defined in [RFC4944] which has been later improved and extended by [ID-6lowpan-hc]. 6LoWPAN compression is stateless, and thus very simple and reliable. It relies on shared information known by all nodes from their participation in that LoWPAN, and the hierarchical IPv6 address space which allows IPv6 addresses to be elided completely most of the time.

The LoWPAN header consists of a dispatch value identifying the type of header, followed by an IPv6 header compression byte indicating which fields are compressed, and then any in-line IPv6 fields. If, for example, UDP or IPv6 extension headers follow IPv6, then these headers may also be compressed using what is called next-header compression [ID-6lowpan-hc]. An example of 6LoWPAN compression is given in Figure 1.10. In the upper packet a one-byte LoWPAN dispatch value is included to indicate full IPv6 over IEEE 802.15.4. Figure 1.11 gives an example of 6LoWPAN/UDP in its simplest form (equivalent to the lower packet in Figure 1.10), with a dispatch value and IPv6 header compression (LOWPAN_IPHC) as per [ID-6lowpan-hc] (2 bytes), all IPv6 fields compressed, then followed by a UDP next-header compression byte (LOWPAN_NHC) with compressed source and destination port fields and the UDP checksum (4 bytes). Therefore in the likely best case the 6LoWPAN/UDP header is just 6 bytes in length. By comparison a standard IPv6/UDP header is 48 bytes in length as shown in Figure 1.12. Considering that in the worst case IEEE 802.15.4 has only 72 bytes of payload available after link-layer headers, compression is important. The 6LoWPAN format and features are described in detail in Chapter 2. Note: these figures showing packet formats are in *box notation*, see Section A.1 for an explanation.

1.3.5   Bootstrapping

Applications of 6LoWPAN most often involve completely autonomous devices and networks, which must autoconfigure themselves without human intervention. Bootstrapping first needs to be performed by the link layer, in order to enable basic communication between nodes within radio range. Basic link layer configuration usually involves the channel setting, default

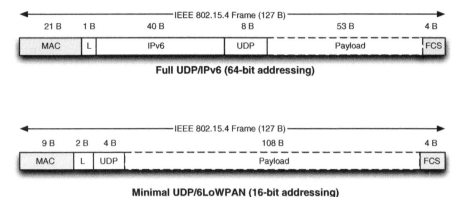

Full UDP/IPv6 (64-bit addressing)

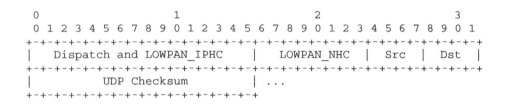

Minimal UDP/6LoWPAN (16-bit addressing)

Figure 1.10 6LoWPAN header compression example (L = LoWPAN header).

```
0                   1                   2                   3
0 1 2 3 4 5 6 7 8 9 0 1 2 3 4 5 6 7 8 9 0 1 2 3 4 5 6 7 8 9 0 1
+-+-+-+-+-+-+-+-+-+-+-+-+-+-+-+-+-+-+-+-+-+-+-+-+-+-+-+-+-+-+-+-+
|   Dispatch and LOWPAN_IPHC    |   LOWPAN_NHC  |  Src  |  Dst  |
+-+-+-+-+-+-+-+-+-+-+-+-+-+-+-+-+-+-+-+-+-+-+-+-+-+-+-+-+-+-+-+-+
|            UDP Checksum        |  ...
+-+-+-+-+-+-+-+-+-+-+-+-+-+-+-+-+
```

Figure 1.11 6LoWPAN/UDP compressed headers (6 bytes).

security key and address settings. Once the link layer is functioning and single-hop communications between devices is possible, 6LoWPAN Neighbor Discovery [ID-6lowpan-nd] is
used to bootstrap the whole LoWPAN.

Neighbor Discovery is a key feature of IPv6, which handles most basic bootstrapping and
maintenance issues between nodes on IPv6 links. Basic IPv6 Neighbor Discovery is specified
in [RFC4861], but is not suitable for use with 6LoWPAN. The 6LoWPAN working group has
defined *6LoWPAN Neighbor Discovery* (6LoWPAN-ND) optimized for low-power wireless
networks and 6LoWPAN in particular [ID-6lowpan-nd]. The 6LoWPAN-ND specification
describes network autoconfiguration and the operation of hosts, routers and edge routers
in LoWPANs. A registry of the nodes in each LoWPAN is kept in the corresponding
edge router, which simplifies IPv6 operation across the network and reduces the amount
of multicast flooding. Additionally 6LoWPAN-ND enables LoWPANs covering many edge
routers connected by a common backbone link (e.g. Ethernet), and the unique generation of
short link-layer addresses. Chapter 3 looks at bootstrapping issues and Neighbor Discovery
in detail. See Appendix A for a reference on basic Neighbor Discovery.

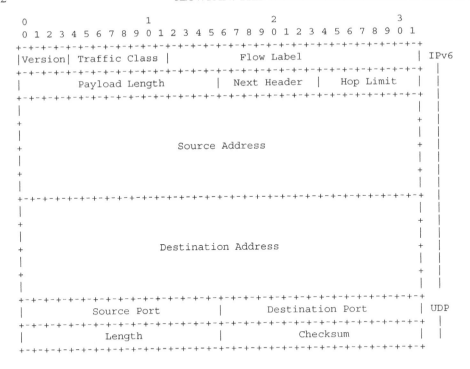

```
 0                   1                   2                   3
 0 1 2 3 4 5 6 7 8 9 0 1 2 3 4 5 6 7 8 9 0 1 2 3 4 5 6 7 8 9 0 1
+-+-+-+-+-+-+-+-+-+-+-+-+-+-+-+-+-+-+-+-+-+-+-+-+-+-+-+-+-+-+-+-+
|Version| Traffic Class |              Flow Label               | IPv6
+-+-+-+-+-+-+-+-+-+-+-+-+-+-+-+-+-+-+-+-+-+-+-+-+-+-+-+-+-+-+-+-+ |
|         Payload Length        |   Next Header  |   Hop Limit   | |
+-+-+-+-+-+-+-+-+-+-+-+-+-+-+-+-+-+-+-+-+-+-+-+-+-+-+-+-+-+-+-+-+ |
|                                                               | |
+                                                               + |
|                                                               | |
+                       Source Address                          + |
|                                                               | |
+                                                               + |
|                                                               | |
+-+-+-+-+-+-+-+-+-+-+-+-+-+-+-+-+-+-+-+-+-+-+-+-+-+-+-+-+-+-+-+-+ |
|                                                               | |
+                                                               + |
|                                                               | |
+                     Destination Address                       + |
|                                                               | |
+                                                               + |
|                                                               | |
+-+-+-+-+-+-+-+-+-+-+-+-+-+-+-+-+-+-+-+-+-+-+-+-+-+-+-+-+-+-+-+-+
|          Source Port          |        Destination Port       | UDP
+-+-+-+-+-+-+-+-+-+-+-+-+-+-+-+-+-+-+-+-+-+-+-+-+-+-+-+-+-+-+-+-+ |
|             Length            |            Checksum            | |
+-+-+-+-+-+-+-+-+-+-+-+-+-+-+-+-+-+-+-+-+-+-+-+-+-+-+-+-+-+-+-+-+
```

Figure 1.12 Standard IPv6/UDP headers (48 bytes).

1.3.6   Mesh topologies

Mesh topologies are common in applications of 6LoWPAN such as automatic meter reading
and environmental monitoring. A mesh topology extends the coverage of the network, and
reduces the cost of needed infrastructure. In order to achieve a mesh topology, multihop
forwarding is required from one node to another. In 6LoWPAN this can be done in
three different ways: link-layer mesh, LoWPAN mesh or IP routing. Link-layer mesh and
LoWPAN mesh are referred to as *Mesh-Under* as the mesh forwarding is transparent to the
Internet Protocol. IP routing is referred to as *Route-Over*.

   Link-layer mesh is possible with some wireless technologies that include multihop
forwarding features such as the recently completed IEEE 802.15.5 standard [IEEE802.15.5].
The original 6LoWPAN specification [RFC4944] includes an option for carrying mesh
source and destination addresses, which can be used by a forwarding algorithm. No standard
algorithms for use with this mesh header have been defined, and therefore the realization of
LoWPAN mesh has been implementation specific. Currently, the most common technique
instead employs IP routing. Routing with 6LoWPAN works just as with standard IP
stacks, an algorithm updates a routing table which IP uses to make next-hop decisions.
The Internet protocol is agnostic to the routing algorithm, and simply forwards packets.
IP routing algorithms for mesh networking are developed in the IETF MANET working

group [MANET] for generic ad hoc networks, and in the IETF ROLL working group [ROLL] specific to wireless embedded applications such as industrial and building automation. IP routing issues and algorithms are discussed in detail in Section 4.2 including information on ROLL.

1.3.7 Internet integration

When connecting a LoWPAN to another IP network or to the Internet, there are several issues to be considered. 6LoWPAN enables IPv6 for simple embedded devices over low-power wireless networks by efficiently compressing headers and simplifying IPv6 requirements. Issues to be considered when integrating LoWPANs with other IP networks include:

Maximum transmission unit: In order to comply with the 1280 byte MTU size requirement of IPv6, 6LoWPAN performs fragmentation and reassembly. Applications designed for the Wireless Embedded Internet should however try to minimize packet sizes if possible. This is to avoid forcing a LoWPAN to fragment IPv6 packets, as this incurs a performance penalty. Additional considerations on fragmentation avoidance are covered in Section 2.7.2.

Application protocols: Application protocols on the Web today depend on payloads of HTML, XML or SOAP carried over HTTP and TCP. This results in payloads ranging in size from hundreds of bytes to several kilobytes. This is far too large for use with 6LoWPAN Nodes. End-to-end application protocols should make use of UDP and compact payload formats (preferably binary) wherever possible, as discussed further in Chapter 5. Technologies which are capable of the transparent compression of web services into a format suitable for 6LoWPAN Nodes are especially interesting.

Firewalls and NATs: In real network deployments firewalls and network address translators (NATs) are a reality. When connecting 6LoWPAN through these there may be several problems that need to be dealt with, for example the blocking of compressed UDP ports and non-standard application protocols used for 6LoWPAN applications, along with the unavailability of static IP addresses.

IPv4 interconnectivity: 6LoWPAN natively supports only IPv6, however often it will be necessary for 6LoWPAN Nodes to interact with IPv4 nodes or across IPv4 networks. There are several ways to deal with IPv4 interconnectivity, including IPv6-in-IPv4 tunneling and address translation. These mechanisms are typically collocated on LoWPAN Edge Routers, on a local gateway router, or on a node configured for that purpose on the Internet. IPv4 interconnectivity is covered in Section 4.3.

Security: When connecting embedded devices to the public Internet, security should always be a major concern as embedded devices are limited in resources and are autonomous. This is very much so with 6LoWPAN as node and network limitations prevent the use of the full IPsec suite, transport layer ("socket") security or the use of sophisticated firewalls on each node. Although link-layer security inside a LoWPAN (employing the 128-bit AES encryption in IEEE 802.15.4) provides some protection, communication beyond LoWPAN Routers is still vulnerable. This increases the need for end-to-end security at the application layer. Security is dealt with further in Section 3.3.

1.4   Network Example

In this section we give a short example of how 6LoWPAN works in practice, concentrating on
the basic things that occur during bootstrapping and operation. Figure 1.13 shows an example
deployment of a Simple LoWPAN, connected through a backhaul link to the IPv6 Internet.
The LoWPAN consists of an edge router, three LoWPAN Routers (R) and three LoWPAN
hosts (H). Additionally there is a remote server on the Internet. This LoWPAN is based on
IEEE 802.15.4 and uses IP routing (which is why there are LoWPAN Routers). Fake IPv6
subnet prefixes and addresses of nodes are included in the figure to make it easy to follow the
example (in reality addresses would be longer).

The router to the Internet advertises the IPv6 prefix 2001:300a::/32 on the backhaul link,
which is used by the edge router for autoconfiguration. The edge router then configures the
IPv6 prefix 2001:300a:1::/48 to its IEEE802.15.4 wireless interface. Note that the LoWPAN
and backhaul link are on different subnets as this uses the Simple LoWPAN model. The
IEEE 802.15.4 wireless devices in the LoWPAN assume a default channel and security
key settings. The edge router starts advertising the IPv6 prefix, which is used by the three
routers to perform Stateless Address Autoconfiguration, and to register with the edge router

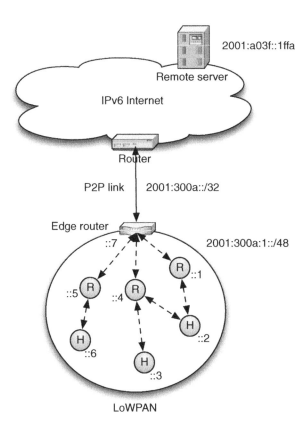

Figure 1.13 A 6LoWPAN example.

using 6LoWPAN-ND. Each LoWPAN Node now has an IPv6 address with a 64-bit IID and additionally receives a generated IPv6 address with a 16-bit IID from the edge router during registration. The IID part of the IPv6 address is shown in the figure as, for example, ::1, which has a full IPv6 address of 2001:300a:1::1. In reality these IIDs would be constructed from 16-bit random numbers. In turn the routers advertise the same prefix to the three hosts, which also register with the edge router. The topology inside the LoWPAN can change freely without affecting the IPv6 addresses of the nodes.

The Neighbor Discovery traffic used for advertising routers and registration is used to initialize the routing algorithm of the routers. Example routes are shown by dashed lines in Figure 1.13. IPv6 source and destination addresses of LoWPAN Nodes are elided during communication. A packet sent to a node on the same link (e.g. ::6 to ::5) does not require in-line IPv6 addresses at all as the link-layer header already contains the source and destination IEEE 802.15.4 addresses. If a packet is forwarded over multiple hops, then just the 16-bit source and destination addresses are carried in-line (e.g. ::3 to ::7) and those addresses are used for routing the packet. Packets destined outside the LoWPAN include either a full IPv6 destination address or a compressed one if compression context for that address is advertised in the LoWPAN. For example, LoWPAN host 2001:300a:1::6 may send a packet to the remote server at 2001:a03f::1ffa. The edge router expands the compressed LoWPAN and IPv6 headers to a full IPv6 header along with the UDP header if compressed. Incoming packets are also processed at the edge router, compressing IPv6 and UDP headers as much as possible.

2

The 6LoWPAN Format

The *Internet Protocol* was designed to enable networking beyond the boundaries of any single network. The Internet is composed of many subnetworks, a number of which are typically traversed by a packet on the way from its source to its destination. For each type of subnetwork, there needs to be an "IP-over-X" specification that defines how to transport IP packets on it. The complexity of such specifications can vary considerably.

At the simple end of the complexity scale for IPv6-over-X specifications is IPv6 over Ethernet [RFC2464], which gets by on little more than four pages of ASCII text (plus boiler-plate), mostly describing packet formats and formats for addresses to be used in these packets. The reason for this simplicity is that the link model used by IPv6 is closely aligned with the capabilities of Ethernet, enabling a very simple one-to-one mapping. In other cases, more work is required to map the services required by the IP layer onto the services provided by the lower layer. The "IP-over-X" specification can amount to a (sub)layer of its own, often called an *adaptation layer*.

One such example is defined by *IPv6 over the point-to-point protocol (PPP)* [RFC5072]. While this specification itself contains only a little more than 10 pages of normative text, it includes by reference the 50-page PPP specification [RFC1661] and, depending on options selected, might invoke specifications as complex as robust header compression (ROHC [RFC3095], more than 150 pages). Finally, PPP itself needs specifications that map it to specific point-to-point links, such as PPP over ISDN (integrated services digital network, a circuit switched digital telephone network based on 64 kbit/s circuits) [RFC1618] or PPP over Ethernet [RFC2516]. Actually, PPP is even more powerful as it can be used as a sublayer for transmitting various protocols over various point-to-point links.

The 6LoWPAN specification for transmitting IPv6 over IEEE 802.15.4 networks, [RFC4944], is not as complex as the PPP set of specifications, but needs to do more work than IPv6 over Ethernet. The services provided by IEEE 802.15.4 are not as powerful as those provided by Ethernet. In addition, the limited performance of IEEE 802.15.4, as well as the desire to limit the amount of energy spent for the actual transmission and reception of data, pose a stronger requirement for optimization in this adaptation layer.

This chapter starts with a short overview of the problems to be solved in 6LoWPAN's adaptation layer, and then describes the 6LoWPAN solutions to each of the problems.

6LoWPAN: The Wireless Embedded Internet   Zach Shelby and Carsten Bormann
© 2009 John Wiley & Sons, Ltd

2.1 Functions of an Adaptation Layer

An "IP-over-X" adaptation layer needs to map IP datagrams to the services provided by the subnetwork, which is usually considered to be at layer 2 (L2) of a layered reference model. A number of problems may need to be solved here:

- Links can be point-to-point or they can provide for the interconnection of multiple IP nodes. When sent on a point-to-point link, a packet is clearly meant to be used by the other node (the one receiving it). For a *multi-access* link, some form of *link-layer addressing* (L2 addressing) is usually provided. Specifically, in a wireless network such as IEEE 802.15.4, a packet may be overheard by multiple receivers, not all of which may need to act on it; the L2 address provides an efficient way to make that decision. Once the IP layer has decided on the IP address of the next hop for a packet, one of the tasks of the adaptation layer is to find out to which link-layer address the packet needs to be addressed to so that it advances on its way to the intended IP-layer destination. This is discussed in Section 2.4. (Interestingly, 6LoWPAN protocols can also take part in actually *assigning* some of the L2 addresses, see Section 3.2.1.)

- The subnetwork may not immediately provide a path for packets to proceed to the next IP node. For instance, when mapping IP to connection-oriented networks such as ISDN or ATM (asynchronous transfer mode, a cell-switched link layer based on virtual circuits of 53-byte equal-size cells), the adaptation layer may need to set up connections (and may have to decide when to close them down again). While LoWPANs are not connection-oriented, in a Mesh-Under situation the adaptation layer may have to figure out the next L2 hop and may need to provide that hop with information about the further direction to forward the packet on. See Section 2.5.1.

- The IP packet needs to be packaged (*encapsulated*) in the subnetwork in such a way that the subnetwork can transport it and the L2 receiver can extract the IP packet again. This leads to a number of subproblems:

  - Links may be able to carry packets of other types than just IP datagrams. Also, there may be a need to distinguish different kinds of encapsulation. Most link layers provide some form of next-layer packet type information, such as the 16-bit ethertype in Ethernet or the PID (protocol ID) in PPP. Not so in IEEE 802.15.4: 6LoWPAN is on its own in identifying different packet encapsulations, as detailed in Section 2.3.

  - IP packets may not fit into the data units that layer 2 can transport. An IP network interface is characterized by the maximum packet size that can be sent using that interface, the MTU (maximum transmission unit). Ethernet interfaces most often have an MTU of 1500 bytes. IPv6 defines a minimum value for the MTU of 1280 bytes, i.e. any maximum packet size imposed by the adaptation layer cannot be smaller than that, and at least 1280 byte or smaller packets have to go through. IEEE 802.15.4 can only transport L2 packets of up to 127 bytes (and a significant part of this can be consumed for L2 purposes). In order to be able to transport larger IPv6 packets, there needs to be a way to carve up the L3 packets and put their contents into multiple L2 packets. The next IP node then needs to put those parts of a packet together again and reconstruct the IP packet. This process is often called segmentation and reassembly; 6LoWPAN calls it *fragmentation* and reassembly in analogy to the IP layer fragmentation process; this is discussed in Section 2.7.

- IP was designed so that each packet stands completely on its own. This leads to a header that may contain a lot of information that could be inferred from its context. In a LoWPAN, the typical IP/UDP header size of 48 bytes already consumes a significant part of the payload space available in a single IEEE 802.15.4 packet, leaving little for applications before fragmentation has to set in. The obvious fix may be to redesign (or avoid the use of) IP. A better approach is to eliminate large parts of the redundancy at the L3–L2 interface, and this has turned out to be a good architectural position to provide *header compression*. Existing IETF standards for header compression (such as ROHC mentioned above) are too heavyweight for LoWPAN Nodes; therefore 6LoWPAN comes with its own header compression, which we will discuss in Section 2.6.

A good source of background information and additional considerations about layer 2 network characteristics and adaptation layers is the IETF informational RFC *Advice for Internet Subnetwork Designers* [RFC3819].

2.2   Assumptions About the Link Layer

The Internet has to work on a rich variety of link-layer technologies, with wildly varying characteristics. In 1998, the IETF recognized that some of these characteristics had performance implications that were not always well understood by the designers of the link layers. In particular, there was and is a tendency for link-layer designers to try to solve as many problems as possible on the link layer, even if these problems are better (or can only be) solved at other layers. The IETF created a working group to investigate the *Performance Implications of Link Characteristics* (PILC), which delivered a number of informational RFCs, including [RFC3819].

IEEE 802.15.4, at 305 densely printed pages (about half of which is just chapter 7, the MAC layer [IEEE802.15.4]), could be cited as an example of the tendency to create rather complex MAC layers, even if the actual technology addressed thrives on its simplicity. Part of the complexity of the document comes from the need to make the standard a part of the IEEE 802 series, which has developed its own reference model and descriptive terminology for physical and MAC layer standards, but a large part stems from the attempt to create consensus by allowing choice from a large number of options. This section discusses the subset of options that are actually being used in 6LoWPAN; Appendix B is provided as a convenient reference for the most important formats and characteristics of IEEE 802.15.4.

2.2.1   Link-layer technologies beyond IEEE 802.15.4

6LoWPAN has been designed with IEEE 802.15.4 in mind. A well-targeted focus on that important link-layer technology was burned into the charter of the 6LoWPAN Working Group and has certainly helped the WG not to wander off into complex, hard to implement generalizations. Certain optional concepts in the 6LoWPAN format specification are closely tied to features of the IEEE 802.15.4 link layer, such as using the PAN ID for address management.

That said, it is important to recognize that the product of the 6LoWPAN WG does exhibit significant generality. Over time, the WG became increasingly reluctant to bind 6LoWPAN to more exotic features of the IEEE 802.15.4 MAC layer. As discussed in Section 1.3.2, 6LoWPAN instead minimizes the functionality that it actually requires from its link layer. But also, the support for IEEE 802.15.4 can be considered to be a lead-in to a wider set of emerging standards: just as Ethernet has shaped other technologies in the link-layer space such as the IEEE 802.11 WLAN standards, there is good reason to expect that new specifications in the wireless embedded space will attempt to stay on a par with the feature set of IEEE 802.15.4, making 6LoWPAN applicable to a much wider set of technologies. First instances of this trend are already visible today, as discussed in Section 1.3.2. In this book, however, we will confine ourselves to the firm ground provided by the well-established IEEE 802.15.4 standard.

2.2.2 Link-layer service model

6LoWPAN attempts to be very modest in its requirements on the link layer. The basic service required of the link layer is for one node to be able to send packets of a limited size to another node within radio reach (i.e. a unicast packet). As always in IP, there is no expectation of reliability; the 6LoWPAN protocols are particularly cognizant of the fact that low-power wireless links cannot always offer a high reliability in packet delivery. In particular, in such a network there is not always a clearly defined boundary of reachability. In wired networks such as Ethernet, it is pretty clear whether a node is plugged in and, if yes, into which Ethernet (link); typically all nodes on an Ethernet can communicate with each other. In a LoWPAN, a node A may be barely (or not at all) in radio range from another node C while both have reasonable error rates to and from a node B.

Instead, 6LoWPAN's requirements on a link are relaxed to an assumption that, with respect to a node A and during a period of time, there is a set of nodes relatively likely to be reachable from A. In this book, we call this set the *one-hop neighborhood* of A. In addition, there is an assumption that A can send a local *broadcast* packet that could be (but is not necessarily always) received by all nodes in A's one-hop neighborhood.

The IEEE 802.15.4 MAC layer defines four types of frames:

Data frames for the transport of actual data, such as IPv6 frames packaged according to the 6LoWPAN format specification;

Acknowledgment frames that are meant to be sent back by a receiver immediately after successful reception of a data frame, if requested by the acknowledgment request bit in the data frame MAC header;

MAC layer command frames, used to enable various MAC layer services such as association to and disassociation from a coordinator, and management of synchronized transmission; and

Beacon frames, used by a coordinator to structure the communication with its associated nodes.

The 6LoWPAN specification only concerns itself with data frames, which are used to carry the protocol data units (PDUs) defined by the LoWPAN adaptation layer, which in turn contain embedded IPv6 packets (or parts of those). In the following text, we will just use the term *6LoWPAN PDU*.

IEEE 802.15.4 data frames may optionally request that they be acknowledged. The 6LoWPAN format specification actually recommends that acknowledgments be requested so that data frames lost on the wireless link can be recovered right at the link layer. IEEE 802.15.4 defines a maximum number of times the sender is supposed to retry a frame after not receiving an acknowledgment, *macMaxFrameRetries*, a number between 0 and 7 that defaults to 3. Such a relatively small *persistence* is also recommended by the "Advice" RFC: on the one hand, it is preferable to resolve packet losses locally within a single hop. On the other hand, persisting for too long is likely to lead to a situation where the link layer is still busy retrying the original frame while upper layers already have decided to retransmit on their own [RFC3819, section 8.1]. (Unfortunately, IEEE 802.15.4's acknowledgment frames are defined in such a way that they cannot make use of data link layer security. An attacker can therefore easily simulate successful reception of data frames that were actually lost. However, this is just another way of getting the overall effect of jamming the wireless signal.)

2.2.3 Link-layer addressing

The link layer must have some concept of globally unique addressing. 6LoWPAN assumes that there is a very low likelihood of two devices coming up in the network with the same link-layer address (as we'll see in Section 3.2, this case is treated as a failure, possibly impeding performance). The fact that an address uniquely identifies a node does not mean that it is by itself useful for locating the node globally, i.e., the link-layer address is not *routable*, and it is not by itself useful for determining if a node is on the same or a different network.

Data frames carry both a source and a destination address. The destination address is used by a receiver to decide whether the frame was actually intended for this receiver or for a different one. The source address is mainly used to look up the keying material for link-layer security, but may also play a role in mesh forwarding (see Section 2.5). IEEE 802.15.4 nodes are permanently identified by EUI-64 identifiers, which weigh in at 8 bytes. As a pair of 64-bit source and destination addresses already consumes one eighth of the usable space in a packet, IEEE 802.15.4 also defines a *short address* format (see Section 2.4). These 16-bit addresses can be dynamically assigned during the bootstrapping of the network, see Section 3.2.1. If a 16-bit address is available in addition to the 64-bit address, some space can be saved in the packet. (For communication from/to a coordinator, IEEE 802.15.4 allows for frames in which either the source or destination addresses are completely elided. However, 6LoWPAN requires that both source and destination addresses be included in the IEEE 802.15.4 frame header.)

IEEE 802.15.4 augments both the source and destination address by a 16-bit *PAN identifier* each. This identifier is meant to separate different IEEE 802.15.4 networks that happen to be in radio range of each other. IEEE 802.15.4 actually stipulates a procedure for resolving PAN identifier conflicts between different coordinators. A much more reliable way to ensure that two 6LoWPAN networks do not accidentally mix up their messages is to use encryption and message integrity checks – the keying material of two networks is extremely likely to be different, so there is no potential for confusion.

2.2.4   Link-layer management and operation

The need to reliably separate different LoWPANs leads to the requirement of data confidentiality and integrity. It is well known that IEEE 802.11 created a significant deployment problem for itself by initially only defining a very weak form of security, until IEEE 802.11i finally fixed this more than half a decade later. IEEE 802.15.4 clearly has learned that lesson and made strong link-layer mechanisms for security an integral part of the standard right from the outset. Section 3.3 discusses how 6LoWPAN may make use of this functionality. One important observation at this point is that the mechanisms provided for encryption and message integrity check including key identification can consume up to 30 bytes of additional space in each data frame. This further reduces the space available for 6LoWPAN PDUs, and it makes the precise amount of space actually available depend on the security parameters chosen.

With respect to the other powerful features of the IEEE 802.15.4 MAC layer, 6LoWPAN strives to maintain a neutral stance. For instance, 6LoWPAN does allow the use of IEEE 802.15.4's *beacon-enabled networks*, although this is not usually being done in practice today. In beacon-enabled mode, a coordinator sends out regular beacon frames. Other devices then synchronize themselves to timeslots defined by those frames, including so-called *superframes* that can provide a contention-free *guaranteed time service* (GTS). Parametrizing and managing all these mechanisms is not easy; see Section 7.1 for an alternate way to achieve similar benefits.

6LoWPAN networks are much more likely to run in *beaconless mode*, performing wireless media access control via IEEE 802.15.4's contention-based channel access method, which IEEE 802.15.4 calls *unslotted CSMA/CA*. When a node wishes to transmit, it first sets a variable *BE* to *macMinBE* (default value: 3). It then waits for a random period in the range 0 to ($2^{BE} - 1$) unit times (where the unit time is 20 symbol periods) and then performs a *clear channel assessment*. If that determines the channel to be idle, the node transmits. If not, it increments BE, up to a value of *macMaxBE* (3–8, by default 5), and waits again. (A node finally gives up after having tried *macMaxCSMABackoffs* times, a value in the range of 0–5, by default 4.) Acknowledgment frames, if requested, are sent with a very short delay (*turnaround time*) after reception without using this CSMA/CA algorithm.

2.3   The Basic 6LoWPAN Format

Appendix B.2 gives an overview over the basic data packet format defined by IEEE 802.15.4. As mentioned in the introduction, this format does not contain any fields that further identify the payload that a data packet is carrying: there is no *multiplexing* information that would allow a receiver to distinguish 6LoWPAN packets from any other data packets that might be sent, or to distinguish between different kinds of 6LoWPAN packets. Therefore, one of the tasks of the 6LoWPAN encapsulation format is to provide a packet type identifier not provided by IEEE 802.15.4 itself.

6LoWPAN does not model its type identifier on any existing identifying schemes such as the 2-byte IEEE 802.3 Ethertype or the 8-byte SNAP (subnetwork access protocol) header often used with IEEE 802.2-based encapsulations [RFC1042]. These type identifiers provide very good long-term extensibility, but were considered too wasteful for the short packets allowed by IEEE 802.15.4. Instead, the first byte of the payload is used as a *dispatch* byte,

Table 2.1 The two most-significant bits in the dispatch byte.

00	Not a LoWPAN packet (NALP)
01	Normal dispatch
10	Mesh header, see Section 2.5 below
11	Fragmentation header, see Section 2.7 below

providing both a type identifier and possibly further information within the subtype. Of the 256 different values possible for the dispatch byte, the 64 values with the two most significant bits set to zero are reserved for IEEE 802.15.4 uses outside of 6LoWPAN (see Section 7.1 for one such use). The other $3 \times 64 = 192$ values are roughly grouped into three classes, also based on the two most significant bits (see Table 2.1). Note that this was an initial allocation; with the further development of 6LoWPAN, additional functions are slowly creeping into the unused spaces. Allocations of dispatch byte values are tracked in an IANA registry, *http://www.iana.org/assignments/6lowpan-parameters/*.

One dispatch byte value, 01000001, is allocated for transporting unmodified IPv6 packets, as suggested in Figure 2.1. This would be the common case on Ethernet (which uses the ethertype 0x86DD for this purpose). On a LoWPAN, however, it is much more likely that a sender would use some form of header compression and/or some of the other functions discussed in this chapter instead. Note that there is no attempt to preserve the nice 32-bit/64-bit field alignment provided by IPv6 – this was deemed unnecessary for the tiny microcontrollers likely to be used in 6LoWPAN devices.

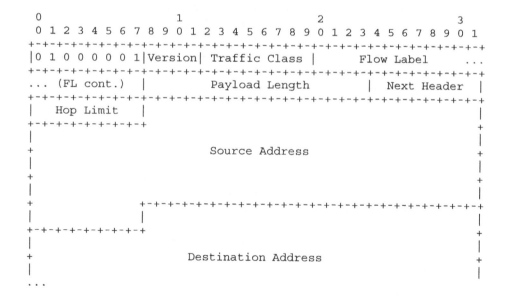

Figure 2.1 Uncompressed IPv6 packet with 6LoWPAN header.

Two other values are allocated out of the 01 space: 01010000=LOWPAN_BC0 to carry a sequence number for duplicate detection in flooding-based broadcasting mechanisms, discussed in Section 2.8, and 01111111=ESC, reserved for extending the range of dispatch values beyond one byte.

Some of the formats defined by 6LoWPAN are designed to carry further 6LoWPAN PDUs as their payload. When multiple headers need to be present, the question is which header should be transported as the payload of which other header, i.e., in which order the headers should be nested. To make this work reliably, 6LoWPAN specifies a well-defined nesting order. If present, the various 6LoWPAN headers should be used in the following order:

Addressing: the mesh header (10nnnnnn, see Section 2.5), carrying L2 original source and final destination addresses and a hop count, followed by a 6LoWPAN PDU;

Hop-by-hop processing: headers that essentially are L2 hop-by-hop options such as the broadcast header (LOWPAN_BC0, 01010000, that carries a sequence number to be checked at each forwarding hop, see Section 2.8), followed by a 6LoWPAN PDU;

Destination processing: the fragmentation header (11nnnnnn, see Section 2.7), carrying fragments that, after possibly having been carried through multiple L2 hops, need to be reassembled to a 6LoWPAN PDU on the destination node;

Payload: headers carrying L3 packets such as IPv6 (01000001, see Figure 2.1), LOW-PAN_HC1 (01000010, see Section 2.6.1), or LOWPAN_IPHC (011nnnnn, see Section 2.6.2).

This is the same order in which the analogous headers at the IPv6 level need to be ordered. The striking difference is that IPv6 has a *next-header* field somewhere in every header that identifies the type of the nested (following) PDU, while 6LoWPAN uses a dispatch byte at the beginning of each PDU to identify its own type. Originally, this has been motivated by the lack of multiplexing information available from the IEEE 802.15.4 MAC layer, but it also was considered to be simpler to process in a constrained implementation. Table 2.2 summarizes the allocations that have been made or are likely to be made soon out of the 256-position space provided by the dispatch byte.

2.4   Addressing

An IP adaptation layer usually involves at least two kinds of addresses: link-layer (L2) addresses and IP (L3) addresses. 6LoWPAN has the additional complication of supporting two address formats at the link layer: 64-bit EUI-64 addresses and dynamically assigned 16-bit short addresses.

As the IEEE encourages for the design of new link layers without a need of direct Ethernet compatibility, IEEE 802.15.4 employs the 64-bit IEEE EUI-64 (extended unique identifier [EUI-64]), a globally unique bit combination which is typically assigned by the manufacturer of a device. To ensure global uniqueness of the EUI-64s, the manufacturer first has to buy a 24-bit OUI (*organizationally unique identifier*) from the IEEE; these are available for a one-time fee of (as of 2009) USD 1650. For each device, the manufacturer then builds the EUI-64 from the OUI and 40 bits of *extension identifier* chosen by the manufacturer, e.g. by

Table 2.2 Current and proposed dispatch byte allocations.

From	To	Allocation
00 000000	00 111111	NALP – Not a LoWPAN frame (NALP)
01 000000		reserved for future use
01 000001		IPv6 – uncompressed IPv6 packets
01 000010		LOWPAN_HC1 – compressed IPv6, see Section 2.6.1
01 000011	01 001111	reserved for future use
01 010000		LOWPAN_BC0 – broadcast, see Section 2.8
01 010001	01 011111	reserved for future use
01 100000	01 111111	proposed for LOWPAN_IPHC, see Section 2.6.2
01 111111		ESC – Additional Dispatch byte follows (preempted by IPHC)
10 000000	10 111111	MESH – Mesh header, see Section 2.5
11 000000	11 000111	FRAG1 – Fragmentation Header (first), see Section 2.7
11 001000	11 011111	reserved for future use
11 100000	11 100111	FRAGN – Fragmentation Header (subsequent), see Section 2.7
11 101000	11 101011	proposed for fragment recovery [ID-thubert-sfr]
11 101100	11 111111	reserved for future use

assigning each device a serial number (or in some other way, the only requirements being uniqueness and avoiding certain reserved values). The procedure is similar to that used for creating Ethernet's 48-bit MAC addresses, except that a 40-bit extension identifier should carry a manufacturer for a long time! The resulting structure of the EUI-64 is illustrated in Figure 2.2.

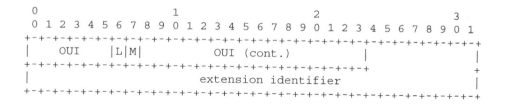

Figure 2.2 Composition of an EUI-64.

Note that two bits out of the OUI field are reserved: the least significant bit (called M here) of the first byte is used to distinguish multicast addresses from unicast ones; the second least significant bit (called L here) is used to distinguish locally assigned addresses from *universal addresses* assigned globally using the OUI/extension scheme. The somewhat peculiar position of these reserved bits in the figure stems from the difference between least-significant-bit-first transmission favored by most IEEE standards and the traditional most-significant-bit-first presentation of packet formats in RFCs. Note that there is little practical effect of the different preferences for bit order within a byte, while the byte order of the

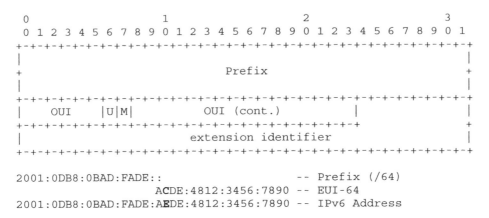

```
 0                   1                   2                   3
 0 1 2 3 4 5 6 7 8 9 0 1 2 3 4 5 6 7 8 9 0 1 2 3 4 5 6 7 8 9 0 1
+-+-+-+-+-+-+-+-+-+-+-+-+-+-+-+-+-+-+-+-+-+-+-+-+-+-+-+-+-+-+-+-+
|                                                               |
+                           Prefix                              +
|                                                               |
+-+-+-+-+-+-+-+-+-+-+-+-+-+-+-+-+-+-+-+-+-+-+-+-+-+-+-+-+-+-+-+-+
|    OUI     |U|M|         OUI (cont.)         |                 |
+-+-+-+-+-+-+-+-+-+-+-+-+-+-+-+-+-+-+-+-+-+-+-+-+                +
|                     extension identifier                      |
+-+-+-+-+-+-+-+-+-+-+-+-+-+-+-+-+-+-+-+-+-+-+-+-+-+-+-+-+-+-+-+-+

2001:0DB8:0BAD:FADE::                  -- Prefix (/64)
               ACDE:4812:3456:7890 -- EUI-64
2001:0DB8:0BAD:FADE:AEDE:4812:3456:7890 -- IPv6 Address
```

Figure 2.3 Composition of an IPv6 address from an EUI-64: U is the inverted L bit.

constituent bytes of larger numbers can have a significant effect. See also the discussion of *box notation* in Section A.1.

One easy way of forming an IPv6 address is to combine a 64-bit prefix with a 64-bit EUI-64 as an interface ID (IID) to yield a 128-bit value. The designers of IPv6 only applied one little twist: they inverted the L bit (turning it into a U bit for universal address) [RFC4291]. This allows the use of easy-to-remember addresses such as 2001:db8::1 for locally assigned addresses. Talking in hexadecimal IPv6 address representations, universal addresses have the first 16-bit subfield in the interface ID XORed with 0200 hexadecimal. The resulting format of an IPv6 address built from an EUI-64 is illustrated in Figure 2.3, note how ACDE changes into AEDE.

For many link layers, this way of assigning IPv6 addresses is just a convention, designed to minimize the probability of a collision occurring in *IPv6 Stateless Address Autoconfiguration* [RFC4862]. In 6LoWPAN, this derivation of IPv6 addresses from link-layer addresses is *mandatory*, as the 6LoWPAN optimizations to the Neighbor Discovery protocol rely on this mapping – see Section 3.2. Note that this means that the *privacy extensions for Stateless Address Autoconfiguration in IPv6* [RFC4941] cannot be used in 6LoWPAN. To a certain extent, this is compensated by the ability to use dynamically assigned 16-bit *short addresses* instead of 64-bit addresses in interface IDs.

In IEEE 802.15.4, the assumption is that 16-bit short addresses are assigned by the PAN coordinator during the association procedure. While this is still possible if that part of the IEEE 802.15.4 MAC is in use, the more likely administrator of 16-bit short addresses is the edge router according to the 6LoWPAN Neighbor Discovery procedures (see Section 3.2). However they are being assigned, the address space for 16-bit short addresses is partitioned by 6LoWPAN into four classes, as detailed in Table 2.3.

To form an IPv6 address out of a 16-bit short address, the short address, if it was assigned by a PAN coordinator, is combined with a PAN identifier as shown in Figure 2.4. Note that the rule for the universal/local bit requires bit 6 of the PAN identifier to be zero, as this is not an IEEE assigned universal EUI-64, so the required zero bit cuts right into the PAN identifier as illustrated. For the more likely case that 16-bit short addresses are assigned by the edge

Table 2.3 Address ranges for 16-bit short addresses.

0xxxxxxxxxxxxxxx	Available for assignment as unicast address
100xxxxxxxxxxxxx	Multicast address (see Section 2.8)
1010000000000000 to	
1111111111111101	Reserved by 6LoWPAN
111111111111111x	0xFFFE and 0xFFFF are reserved by IEEE 802.15.4

router using the Neighbor Discovery extensions, or, more generally, "if no PAN ID is known" [RFC4944, section 6], the PAN identifier can be replaced by 16 zero bits, which also satisfies the rule for a zero universal bit.

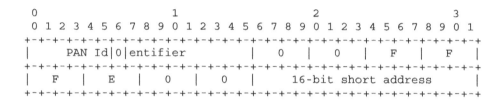

```
 0                   1                   2                   3
 0 1 2 3 4 5 6 7 8 9 0 1 2 3 4 5 6 7 8 9 0 1 2 3 4 5 6 7 8 9 0 1
+-+-+-+-+-+-+-+-+-+-+-+-+-+-+-+-+-+-+-+-+-+-+-+-+-+-+-+-+-+-+-+-+
|      PAN Id|0|entifier       |   0   |   0   |   F   |   F   |
+-+-+-+-+-+-+-+-+-+-+-+-+-+-+-+-+-+-+-+-+-+-+-+-+-+-+-+-+-+-+-+-+
|   F   |   E   |   0   |   0   |      16-bit short address     |
+-+-+-+-+-+-+-+-+-+-+-+-+-+-+-+-+-+-+-+-+-+-+-+-+-+-+-+-+-+-+-+-+
```

Figure 2.4 Interface identifier for 16-bit short addresses.

2.5 Forwarding and Routing

Packets will often have to traverse multiple radio hops on their way through the LoWPAN. This involves two related processes: *forwarding* and *routing*. Both can be performed at layer 2 or at layer 3. Routing typically involves one or more routing protocols, which will be discussed in Section 4.2. In each node, the routing protocol fills in a *routing information base* (RIB), which contains all the information needed to run the routing protocol. The RIB usually can be simplified to a *forwarding information base* (FIB), which is consulted when a packet arrives that needs to be forwarded. Some routing protocols fill the FIB *proactively*, i.e., the FIB should always contain an entry for each packet that can actually be forwarded, while others operate *reactively* by filling in gaps in the FIB only as packets arrive.

Figure 2.5 shows the usual illustration for routing at the network layer: packets are sent through some link and arrive at a router on an interface *if0*. The router looks up the destination address in its FIB, selects an interface to forward it out on (here: *if1*), together with a link-layer address to send it to, and sends the packet encapsulated with the new link-layer address out via that interface.

In a LoWPAN, forwarding is not motivated by using two different instances of the link layer (links), but by the fact that the first node may not have the radio range to reach the third node. So the interface that the packet arrives on at the router node is usually the same interface that is used again for sending it out (sometimes this single-interface configuration

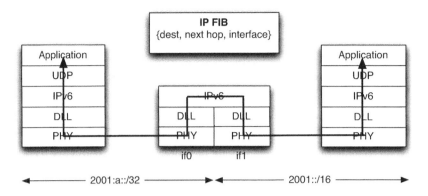

Figure 2.5 The IP routing model.

has been called "router on a stick", except that here the stick is wireless and thus just an antenna is visible), see Figure 2.6.

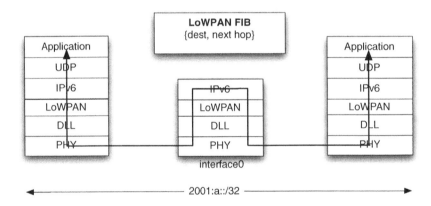

Figure 2.6 The LoWPAN routing model (L3 routing, "Route-Over").

The illustrations in Figures 2.5 and 2.6 show the actual routing on the IP level. Routing and forwarding in a LoWPAN can happen both below the IP layer and on the IP layer. The next two sections consider the implications on the format level.

2.5.1 L2 forwarding ("Mesh-Under")

When routing and forwarding happen at layer 2, they are performed based on layer-2 addresses, i.e., 64-bit EUI-64 or 16-bit short addresses. The IETF is not usually working on layer-2 routing ("mesh routing") protocols. ISA100 (see Section 7.1) defines one such routing protocol, together with some extensions to the data link layer that make the fact that routing and forwarding is happening at layer 2 essentially invisible to the LoWPAN adaptation layer, as illustrated in Figure 2.7. The situation might be similar when using an IEEE-defined mesh protocol for IEEE 802.15.4, such as [IEEE802.15.5].

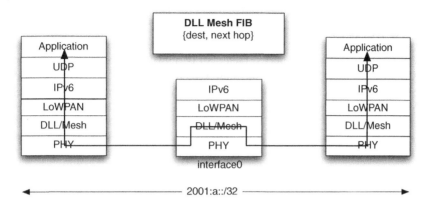

Figure 2.7 DLL mesh forwarding below the LoWPAN adaptation layer.

In case the actual link-layer forwarding is not hidden from the LoWPAN adaptation layer, as illustrated in Figure 2.8, there is one problem to solve: the link-layer headers describe the source and destination addresses for the current layer-2 hop. To forward the packet to its eventual layer-2 destination, the node needs to know its address, the *final destination address*. Also, to perform a number of services including reassembly, nodes need to know the address of the original layer-2 source, the *originator address*. Since each forwarding step overwrites the link-layer destination address by the address of the next hop and the link-layer source address by the address of the node doing the forwarding, this information needs to be stored somewhere else. 6LoWPAN defines the *mesh header* for this.

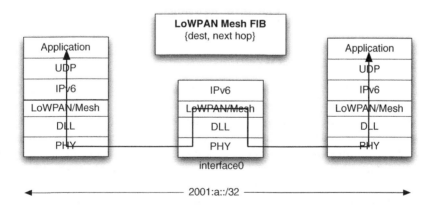

Figure 2.8 LoWPAN adaptation layer mesh forwarding.

In addition to the addresses, the mesh header stores a layer-2 equivalent of an IPv6 Hop Limit. Since the diameters of useful wireless multihop networks are usually small, the format is optimized for *hops left* values below 15 by only allocating four bits to that value (first case in Figure 2.9). If a value of 15 or larger is needed, the 6LoWPAN packet encoder needs to insert an extension byte (second case in Figure 2.9). This value must be decremented by a forwarding node before sending the packet on its next hop; if the value reaches zero, the packet is discarded silently. (Note that the lack of any error message means that the *traceroute* functionality that was one of the success factors of IP networking cannot be implemented for 6LoWPAN Mesh-Under.) In an implementation, there is no need to remove the extension byte when the hops left value drops to 14 or less by the decrementing process; the packet can be sent on as is or it can be optimized by removing the byte and moving the value into the dispatch byte.

```
 0                   1                   2                   3
 0 1 2 3 4 5 6 7 8 9 0 1 2 3 4 5 6 7 8 9 0 1 2 3 4 5 6 7 8 9 0 1
+-+-+-+-+-+-+-+-+-+-+-+-+-+-+-+-+-+-+-+-+-+-+-+-+-+-+-+-+-+-+-+-+
|1 0|V|F|HopsLft| originator address, final address ...
+-+-+-+-+-+-+-+-+-+-+-+-+-+-+-+-+-+-+-+-+-+-+-+-+-+-+-+-+-+-+-+-+

+-+-+-+-+-+-+-+-+-+-+-+-+-+-+-+-+-+-+-+-+-+-+-+-+-+-+-+-+-+-+-+-+
|1 0|V|F|1 1 1 1|  Hops Left  | originator address, final address...
+-+-+-+-+-+-+-+-+-+-+-+-+-+-+-+-+-+-+-+-+-+-+-+-+-+-+-+-+-+-+-+-+
 \_ dispatch  _/
```

Figure 2.9 Mesh addressing type and header.

The V and F bits indicate whether the originator (or "very first") address and the final Destination address, respectively are 16-bit short (1) or 64-bit EUI-64 (0) addresses.

The mesh header, if used, must be the first header in a 6LoWPAN header stack, as it carries the addresses for forwarding. The next header might be a LOWPAN_BC0 header for multicasting, see Section 2.8. Further headers in a stack might include fragmentation headers, which are not touched during mesh forwarding – fragments are mesh-forwarded independently of each other.

2.5.2   L3 routing ("Route-Over")

Layer-3 Route-Over forwarding is illustrated in Figure 2.6. In contrast to layer-2 mesh forwarding, layer-3 Route-Over forwarding does not require any special support from the adaptation layer format. Before the layer-3 forwarding engine sees the packet, the adaptation layer has done its work and decapsulated the packet – at least conceptually (implementations may be able to perform some optimizations by keeping the encapsulated form if they know how to rewrite it into the proper encapsulated form for the next layer-3 hop).

Note that this in particular means that fragmentation and reassembly are performed at each hop in Route-Over forwarding – it is hard to imagine otherwise, as the layer-3 addresses are part of the initial bytes of the IPv6 header, which is present only in the first fragment of a larger packet. Again, implementations may be able to optimize this process by keeping virtual reassembly buffers that remember just the IPv6 header including the relevant addresses (and the contents of any fragments that arrived out of order before the addresses).

2.6 Header Compression

An important characteristic of 6LoWPAN is the limited payload size of packets provided by IEEE 802.15.4, about half of which would be consumed by the size of an IPv6 header already. While larger packets can be sent using fragmentation and reassembly (see Section 2.7), 6LoWPAN is most efficient when all IPv6 packets can be made to fit into a single IEEE 802.15.4 packet. Even if payloads are small enough to make fragmentation less of a concern, the cost in battery lifetime and in utilization of the rather limited channel capacity of IEEE 802.15.4 networks for the transmission of large IPv6 headers has often been cited as a reason for not wanting to run IP on wireless embedded networks.

Much of the information that is repeatedly sent in sequences of network headers is *redundant*, i.e. could be compressed away. For instance, on the last hop to a host, the values for the IP destination address field will come from a very small set and therefore could be expressed by a couple of bits instead of the entire 128 bits sent in a full IPv6 header – assuming the last-hop sender and the host have reached agreement on what these bits mean.

Data compression techniques such as *gzip* [RFC1952] and its underlying *DEFLATE* algorithm [RFC1951] are optimized for eliminating redundancies in a given data item or stream. They don't work very well on small data items such as single packets, and applying them on sequences of packets is difficult as packets may be lost, destroying information that would be needed for decoding subsequent packets in the packet sequence.

Data compression algorithms are best applied to the application layer content of the packets. A separate set of techniques has evolved that focuses on compressing the stacks of headers in sequences of packets: *header compression*.

Header compression can be performed end-to-end, but is then limited to compressing the headers that are within the payload of the IP header, as the routers on the way between compressor and decompressor still have to see full IP headers. Since the largest header in many IPv6 header stacks is the IP header itself, this is not very efficient. Instead, most header compression schemes operate *hop-by-hop*, i.e. as part of the adaptation layer. This allows compressing the full header stack including the IP header immediately before sending the packet on a link, and decompressing and thus reconstructing the header stack in full before the packet is possibly routed and sent on via a different link, quite likely with a different header compression scheme or at least different header compression parameters (see Figure 2.10). The beauty of this approach is that deploying header compression becomes a local decision between two neighbors: only the two nodes on the ends of a link need to agree on its use (and the specific parameters to be used).

There is a significant body of RFCs dealing with header compression: from 1990's first generation *Van Jacobson header compression* [RFC1144], via the second-generation *IP header compression* [RFC2507] and its extension to RTP [RFC2508, RFC2509, RFC3544, RFC3545], to the third-generation *ROHC (robust header compression)* family of standards [RFC3095, RFC3241, RFC3843, RFC4815, RFC4995, RFC4996, RFC5225], which has even developed its own formal notation [RFC4997].

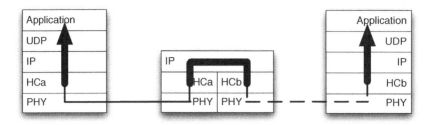

Figure 2.10   Hop-by-hop header compression with two different header compression methods (HCa, thin line; HCb, thin dashed line).

The ROHC standards and its predecessors focus on compressing *flows* of packets, such as the sequence of packets from a single TCP connection or an RTP stream. They expend considerable complexity for compressing the full stack of headers (IP/TCP or IP/UDP/RTP, possibly with embedded extension headers) down to a very small, usually single-digit number of bytes. This works by setting up flow state for each new connection/stream, and stepping through a set of compression states during the exchange of the initial packets of the flow until a high level of compression is attainable.

The main problem that flow-based header compression needs to solve is that the per-flow state, the *context*, is likely to get out of sync between the sender and the receiver when packets get lost. Various techniques such as positive acknowledgments or optimistic compression with checksums and negative acknowledgments either avoid progressing the context state in this case or allow quick recovery (*robustness*).

While significant efficiencies can be reaped for long-running flows such as Voice over IP, the required complexity to achieve these gains ranks ROHC among the top dozen most complex IETF protocols just judging from specification size, in the same league with SIP [RFC3162], NFSv4 [RFC3530], IOTP [RFC2801], iSCSI [RFC3720], IPP [RFC2911], OSPF [RFC2328] etc. A protocol of this complexity was judged inappropriate for the resource-constrained systems envisaged in a LoWPAN.

Instead, the original 6LoWPAN format standard exclusively employs *stateless* compression, see Section 2.6.1. Without state, there is no synchronization problem, and the algorithms become very simple.

The largest part of IPv6 headers are the two 16-byte IPv6 addresses in the header, which together can consume 40 percent or more of the usable space in a 6LoWPAN packet. The original 6LoWPAN format specification defines how to compress certain forms of link-local addresses, but still requires the transmission of globally routable addresses (or at least their 64-bit prefixes) at their full size. Since the whole point of running IPv6 in a LoWPAN is to enable communication beyond the local link, this is unsatisfactory.

Unfortunately, there is no way to compress the large globally routable IPv6 addresses without at least some state. The 6LoWPAN WG has therefore agreed to standardize a second header compression method, *context-based* header compression with a slowly evolving global context, discussed in Section 2.6.2, with a view to deprecating the stateless-only header compression scheme once context-based compression is completed.

2.6.1   Stateless header compression

The 6LoWPAN format specification [RFC4944] defines two header compression schemes
that are designed to work together: HC1 to compress IPv6 headers, and HC2 to compress
UDP headers. HC2 can optionally be used in packets that also use HC1. HC1 is selected
by using a dispatch byte of LOWPAN_HC1 (01000010). The next byte then selects various
options; the final bit, if set, indicates that HC2 is used as well, using another byte of options
(most of which are reserved). Figure 2.11 shows the initial bytes of an HC1- or HC1/HC2-
compressed payload.

```
 0                   1                   2                   3
 0 1 2 3 4 5 6 7 8 9 0 1 2 3 4 5 6 7 8 9 0 1 2 3 4 5 6 7 8 9 0 1
+-+-+-+-+-+-+-+-+-+-+-+-+-+-+-+-+-+-+-+-+-+-+-+-+-+-+-+-+-+-+-+-+
|0 1 0 0 0 0 1 0|SAE|DAE|C|NH |0|       Non-Compressed fields...
+-+-+-+-+-+-+-+-+-+-+-+-+-+-+-+-+-+-+-+-+-+-+-+-+-+-+-+-+-+-+-+-+
 \_ dispatch  _/ \_ HC1 header_/

 0                   1                   2                   3
 0 1 2 3 4 5 6 7 8 9 0 1 2 3 4 5 6 7 8 9 0 1 2 3 4 5 6 7 8 9 0 1
+-+-+-+-+-+-+-+-+-+-+-+-+-+-+-+-+-+-+-+-+-+-+-+-+-+-+-+-+-+-+-+-+
|0 1 0 0 0 0 1 0|SAE|DAE|C|NH |1|S|D|L|_____| N.-C. fields...
+-+-+-+-+-+-+-+-+-+-+-+-+-+-+-+-+-+-+-+-+-+-+-+-+-+-+-+-+-+-+-+-+
 \_ dispatch  _/ \_ HC1 header_/ \_ HC2 header_/
```

Figure 2.11 HC1-compressed IPv6 packet: without and with HC2.

The objective of HC1/HC2 was to enable header compression in an entirely stateless
fashion. In other words, there is no requirement for previous agreement between nodes
exchanging the compressed packets. So HC1/HC2 can essentially just exploit internal
redundancies in the packet, or possibly encode more likely variations of the packet in
fewer bits.

The most useful redundancy in a 6LoWPAN packet is caused by the way IP addresses
are formed from layer-2 addresses, both of which are present in the overall layer-2 packet.
When a 6LoWPAN host sends a packet, the layer-2 source address of the packet will reflect
the MAC address of the host. As discussed in Section 2.4 above, the IID half of the IP
address is created from the MAC address as well, so it is redundant and can be elided. Similar
considerations apply to the two destination addresses present in packets sent to hosts.

It is much harder to compress the first 64 bits of an IP address. HC1 only optimizes the case
where these bits indicate a link-local address, prefix FE80::/64. HC1 allows the compressor
to select elision of each half of both source and destination address independently by using
two bits in the HC1 header each, called SAE (source address encoding) and DAE (destination
address encoding) in Figure 2.11, encoded as shown in Table 2.4.

The rest of the HC1 header is concerned with the compression of the non-address
components of an IPv6 header, which are illustrated in Figure 2.12.

- The version number is obviously always 6 and therefore never needs to be sent.

- Two fields that often are zero in IPv6 headers are the traffic class and the flow label.
  The C bit in the HC1 header, if set, indicates that these bits indeed are zero and are

Table 2.4 HC1 SAE and DAE values.

SAE or DAE value	Prefix	IID
00	sent in-line	sent in-line
01	sent in-line	elided and derived from L2 or mesh address
10	elided and assumed to be link-local (FE80::/64)	sent in-line
11	elided and assumed to be link-local (FE80::/64)	elided and derived from L2 or mesh address

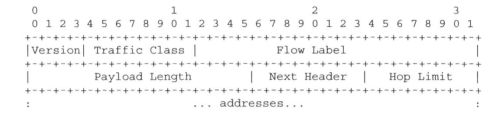

```
 0                   1                   2                   3
 0 1 2 3 4 5 6 7 8 9 0 1 2 3 4 5 6 7 8 9 0 1 2 3 4 5 6 7 8 9 0 1
+-+-+-+-+-+-+-+-+-+-+-+-+-+-+-+-+-+-+-+-+-+-+-+-+-+-+-+-+-+-+-+-+
|Version| Traffic Class |            Flow Label                 |
+-+-+-+-+-+-+-+-+-+-+-+-+-+-+-+-+-+-+-+-+-+-+-+-+-+-+-+-+-+-+-+-+
|         Payload Length        |  Next Header  |   Hop Limit   |
+-+-+-+-+-+-+-+-+-+-+-+-+-+-+-+-+-+-+-+-+-+-+-+-+-+-+-+-+-+-+-+-+
:                       ... addresses...                        :
```

Figure 2.12 IPv6 header: non-address fields.

Table 2.5 HC1 NH values.

00	Next header sent in-line
01	Next header = 17 (UDP)
10	Next header = 1 (ICMP)
11	Next header = 6 (TCP)

not sent. If the C bit is clear, they are included among the non-compressed fields in the sequence explained below.

- The payload length can be inferred from the remaining length of the 6LoWPAN PDU (which in turn can be found out from the link-layer frame or from the fragmentation mechanism) and therefore is never sent in a compressed header.

- The next header field is one full byte, but has a number of values that are much more likely than others. The NH bits in the HC1 header, if non-zero, indicate that the next header field is implied by their value (see Table 2.5; if zero, the next header value is sent in-line).

- Finally, the Hop Limit was considered to be too difficult to compress and therefore is always sent in-line in the non-compressed fields.

The non-compressed fields that follow the HC1 or HC1/HC2 header always start with the Hop Limit (8 bits). After that follow any in-line fields in the same order used for the fields in the HC1 header:

- source address prefix (64 bits), if the high-order bit of SAE is zero;

- source address interface identifier (64 bits), if the low-order bit of SAE is zero;

- destination address prefix (64 bits), if the high-order bit of DAE is zero;

- destination address interface identifier (64 bits), if the low-order bit of DAE is zero;

- traffic class (8 bits) and flow label (20 bits), if the C bit is zero;

- next header (8 bits), if NH is zero;

- any non-compressed fields left by HC2, if present;

- any next headers and payloads not further subject to compression.

The presence of an HC2 header is indicated by the rightmost bit of the HC1 header; this is only fully defined if the NH bits indicate a next header of 17 (UDP). The HC2 header has three bits to indicate the compression of the source port, the destination port, and the length; the UDP checksum is never compressed. It is trivial to compress the length (it can just be inferred by the number of payload bytes remaining) – the L bit that, if set, indicates this is being done might as well always be set.

It is much harder to compress the source and destination port, as this would require the identification of some port numbers that are more likely than others. Since favorite port numbers might differ between applications of 6LoWPAN, the format specification simply favors an arbitrarily chosen subspace of the dynamic port numbers. Any port number between 61616 and 61631 (0xF0Bn) can be compressed to just the lower four bits by setting the S and D bit in the HC2 header for source and destination, respectively, saving three bytes if both bits are set.

Any uncompressed UDP fields follow the other non-compressed fields in the LOW-PAN_HC1 packet in the order they occur in the UDP header (source port, destination port, length, checksum).

Unfortunately, the 6LoWPAN format specification does not explain how non-compressed fields that are not a multiple of eight bits are intended to be handled; this is a problem with the 20-bit flow label as well as the compressed UDP ports (unless both UDP ports are compressed). As implementers are very keen to replace stateless header compression with context-based header compression, it is not clear whether this specification gap will ever be filled.

An example of the best case of stateless header compression is given below: when two adjacent nodes interchange a UDP packet between their link-local addresses, using 0xF0Bn ports numbers, the resulting compressed packet might look like Figure 2.13.

2.6.2 Context-based header compression

While the example given for best-case stateless header compression is impressive, it applies only to a rather unlikely case. In many LoWPANs, most packets will go from LoWPAN hosts

```
0                   1                   2                   3
0 1 2 3 4 5 6 7 8 9 0 1 2 3 4 5 6 7 8 9 0 1 2 3 4 5 6 7 8 9 0 1
+-+-+-+-+-+-+-+-+-+-+-+-+-+-+-+-+-+-+-+-+-+-+-+-+-+-+-+-+-+-+-+-+
|0 1 0 0 0 0 1 0|1 1|1 1|1|0 1|1|1|1|1|_____|  Hop Limit   |
+-+-+-+-+-+-+-+-+-+-+-+-+-+-+-+-+-+-+-+-+-+-+-+-+-+-+-+-+-+-+-+-+
| Sport | Dport |          UDP Checksum         | UDP Data...
+-+-+-+-+-+-+-+-+-+-+-+-+-+-+-+-+-+-+-+-+-+-+-+-+-+-+-+-+-+-+-+-+

|   LOWPAN_HC1   |SAE|DAE|C|NH |1|S|D|L|_____| N.-C. fields...
\_  dispatch   _/ \_ HC1 header_/ \_ HC2 header_/
```

Figure 2.13 Best-case HC1-/HC2-compressed IPv6 packet.

to some nodes external to the LoWPAN or in the inverse direction. Both cases cannot work with link-local addresses; both source and destination addresses will be routable addresses (globally routable or routable within the network of an organization). Generally, only eight of the 32 bytes of IPv6 addresses in such packets will be compressible (for an outward-bound packet, only the IID part of the source address).

After some initial experience with 6LoWPAN and its header compression mechanisms, a second-generation header compression specification emerged. Currently at the stage of a Working-Group Internet-Draft [ID-6lowpan-hc], it has already garnered wide consensus within the WG and is likely to be accepted as a standards-track RFC soon (likely with very few changes, as it has also been adopted by ISA100).

In order to enable the compression of global addresses, this new specification assumes that there is a way for nodes to establish some additional *context* when joining the LoWPAN. This context can then be referred to in 6LoWPAN packets exchanged between the nodes of the LoWPAN and used for address field compression. The new context-based header compression scheme does not define how the context is acquired, but the assumption is that this will be done using 6LoWPAN's extensions to Neighbor Discovery (6LoWPAN-ND, see Section 3.2) or some system standard such as ISA100 (see Section 7.1).

Obviously, the most important correctness aspect about a context-based header compression scheme is making sure that the context stays synchronized between compressor and decompressor. In a wireless network with changing node reachability and sleeping nodes, changes to the global context do not necessarily reach every node immediately. Therefore, the context is divided into multiple slots that may be changed independently. A sending node should only start using a context slot when there is sufficient reason to believe that the updated value for this slot has percolated to the receiving node. When a node might compress a packet still making use of a previous value of a recently changed context slot, ambiguity results in the decompressor. Old values of a context slot should therefore be taken out of use for a while before new values are assigned to this specific context slot. Compressors must stop using context information for a specific context slot when new information appears that does not assign a value to this context slot, while decompressors may want to hold on to the old value for some more time.

To further reduce the probability of damage, LoWPAN Nodes should use context-based header compression only when a higher-layer protocol is in use that protects the IPv6 addresses using some form of pseudo-header-based checksum and/or authenticator, such as UDP, TCP or some application-specific integrity protocol.

Similar to stateless header compression, the context-based header compression is divided into a compression scheme for the IP header (LOWPAN_IPHC) and an optional compression scheme for the next header (LOWPAN_NHC).

```
0                   1                   2                   3
0 1 2 3 4 5 6 7 8 9 0 1 2 3 4 5 6 7 8 9 0 1 2 3 4 5 6 7 8 9 0 1
+-+-+-+-+-+-+-+-+-+-+-+-+-+-+-+-+-+-+-+-+-+-+-+-+-+-+-+-+-+-+-+-+
|0|1|1| TF|N|HLM|C|S|SAM|M|D|DAM| SCI  |  DCI  | ...fields...
+-+-+-+-+-+-+-+-+-+-+-+-+-+-+-+-+-+-+-+-+-+-+-+-+-+-+-+-+-+-+-+-+
 \_ dispatch  _/                    \_if C is set_/
        \_    IPHC base header   _/
```

Figure 2.14 LOWPAN_IPHC header.

The LOWPAN_IPHC base header needs 13 bits. To enable a compact encoding for this likely very frequent header, five of these bits are put into the rightmost bits of the dispatch type; i.e., 32 of the 64 dispatch values in what we called the normal dispatch space are consumed by the header compression scheme. The eight remaining bits are put into one additional byte for the base header, making the total overhead for indicating the packet type and setting the compression parameters two bytes.

The LOWPAN_IPHC header is illustrated in Figure 2.14. The C bit (called CID in the specification [ID-6lowpan-hc]) controls whether a third byte is added to specify Context IDs for the source and destination address. If C is not set, the "default context" is used for both (i.e., the context with context ID = 0).

After the two or three bytes (including the dispatch bytes) of the LOWPAN_IPHC header, all the (parts of) IPv6 header fields follow that have not been compressed away and need to be sent in-line, in the same order as they do in the uncompressed IPv6 header. (Processing the fields in order simplifies implementation appreciably.) This completes the compressed IPv6 header, which is then followed by the next header (e.g., UDP); if the N bit is set, this header is again compressed using LOWPAN_NHC (see below).

From the IPv6 header, the version field is always elided (a value of 6 is understood), as is the IPv6 payload length field, which is inferred from the length of the 6LoWPAN PDU, adjusted for the effects of compression. As in stateless compression, this requires that a valid value for the length of the 6LoWPAN PDU is available; it might come directly from the IEEE 802.15.4 header or, for IPv6 packets fragmented in the adaptation layer, the packet size of the reassembled packet will come from the 6LoWPAN fragmentation header.

The LOWPAN_IPHC header mostly consists of flags and two-bit selectors that control which IPv6 header fields are compressed in which way:

- The TF bits control how the IPv6 header fields traffic class and flow label are handled (Figure 2.15). As the flow label is an unstructured 20-bit label [RFC2460, RFC3697], provision is only made to completely elide it if all bits are zero (the value for packets that are not part of any specific flow, a very likely case for today's traffic). The IPv6 traffic class field is structured into a 6-bit differentiated services control point (DSCP

[RFC2474, RFC3260]) that may be used to group traffic into different classes and two bits used for explicit congestion notification (ECN [RFC3168]), which allow routers to indicate congestion by more subtle means than by dropping a complete packet. In contrast to the previous header compression format, special care has been taken to support the IPv6 header bits for ECN efficiently while compressing away the flow label or the DSCP; the assumption is that ECN and differentiated services can be put to good use in resource-constrained LoWPANs. The three (sub)fields can be sent essentially unchanged (slightly reordered so that ECN is always sent first if sent at all, TF = 00), the DSCP part of the traffic class can be elided if all bits are zero (TF = 01), the flow label can be elided if all of its bits are zero (TF = 10), or both traffic class and flow label can be completely elided if they are both entirely zero (TF = 11).

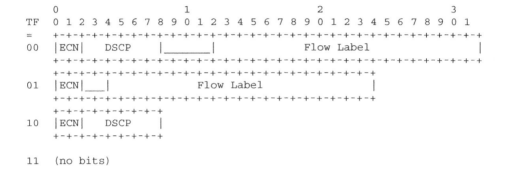

Figure 2.15 LOWPAN_IPHC traffic class and flow label compression.

- As mentioned above, the N bit (called NH in [ID-6lowpan-hc]) controls both whether a next header field is sent in-line and whether the next header uses LOWPAN_NHC.

- The IPv6 header field Hop Limit cannot be completely elided as each router on the way needs to reduce its value. However, it is possible to identify more likely values and use fewer bits for these, at the expense of needing more bits for the less likely cases. Using very efficient schemes such as Huffman coding would require lots of code and make it hard to achieve byte alignment. Instead, the Hop Limit field is compressed by defining three preferred values: 1, 64, and 255. A two-bit field (called HLIM in [ID-6lowpan-hc] and HLM in Figure 2.14) can select one of the three preferred values or, when set to 00, indicate that a separate byte sent in-line will contain the actual value of the Hop Limit field.

In total, Hop Limit compression saves six bits per packet when one of the preferred values is used and costs two bits otherwise, so it is only beneficial if the preferred values occur in at least a third of the cases. A 6LoWPAN Router can possibly assist this and save a byte by reducing the Hop Limit by more than one to meet one of the preferred values; for instance, if a packet with a Hop Limit of 70 arrives at an edge router, the edge router might reduce the Hop Limit to the preferred value of 64 before

Table 2.6 S/SAM values, and D/DAM values when M = 0.

S/SAM	Inline	
0 00	128	inline (address fully inline, no compression)
0 01	64	FE80:0:0:0:inline (link-local + 64-bit Interface ID)
0 10	16	FE80:0:0:0:0:0:0:inline (link-local + 16-bit short address)
0 11	0	FE80:0:0:0:link-layer (link-local + link-layer source address)
1 00	—	reserved
1 01	64	context[0..63]:inline (context + 64-bit Interface ID)
1 10	16	context[0..111]:inline (context + 16-bit short address)
1 11	0	context[0..127] (context)

Table 2.7 EID values in LOWPAN_NHC base header.

0 :	IPv6 Hop-by-Hop Options [RFC2460]
1 :	IPv6 Routing [RFC2460]
2 :	IPv6 Fragment [RFC2460]
3 :	IPv6 Destination Options [RFC2460]
4 :	IPv6 Mobility Header [RFC3775]
5 :	(Reserved)
6 :	(Reserved)
7 :	Nested IPv6 Header, LOWPAN_IPHC encoded

sending it on; similarly, the last hop router can always save a byte by reducing the Hop Limit down to 1. Similarly, a packet source can set the Hop Limit to one of the preferred values and save a byte on the first hop; the value 64 was chosen as a likely first-hop Hop Limit.

- The compression of the source and destination addresses is controlled by S/SAM and D/DAM, respectively. A special flag, M, can be set to indicate that the destination address is a multicast address; this case is discussed in Section 2.8. For unicast, the flags are interpreted as follows: The S flag (called SAC in [ID-6lowpan-hc]) and the D flag (DAC) control whether compression of the respective address is context-based or not. The SAM/DAM selectors control submodes that depend on these bits; each of these submodes specifies a number of bits that are carried in-line in sequence with the other uncompressed parts of the IPv6 header; these bits are then combined with fixed bits and/or bits taken from the context to reconstruct the respective address (see Table 2.6).

If the N bit in the LOWPAN_IPHC header is not set, the compressed IPv6 header contains an in-line next header field and is immediately followed by the IPv6 payload. If the N bit is set, a LOWPAN_NHC header follows. The LOWPAN_NHC base header both implies the next header field and provides parameters for the compression of the fields in that next header. The LOWPAN_NHC base header is defined in an extensible way similar to the dispatch byte of the overall format. For now, only a small part of the space is taken, either for the identification and compression of UDP (Figure 2.16, see below), or to indicate certain IPv6 extension headers (slightly adjusted to eliminate some padding), and, by setting another N bit, optionally the presence of further LOWPAN_NHC headers and the elision of the next header field in the extension header (see Table 2.7 and Figure 2.17).

```
0  1  2  3  4  5  6  7
+-+-+-+-+-+-+-+-+
|1|1|1|1|0|C|  P  |
+-+-+-+-+-+-+-+-+
```

Figure 2.16 LOWPAN_NHC base header for UDP.

```
0  1  2  3  4  5  6  7
+-+-+-+-+-+-+-+-+
|1|1|1|0| EID |N|
+-+-+-+-+-+-+-+-+
```

Figure 2.17 LOWPAN_NHC base header for IPv6 extension headers.

Figure 2.16 shows the LOWPAN_NHC base header values for UDP. The P field and C bit control how the port number and checksum fields of the UDP header are compressed. The UDP length is redundant and always elided.

- The C bit controls whether the UDP checksum is elided. If it is set, the UDP checksum is removed by the compressor and must be recomputed at the decompressor. (An implementation may leave out the latter step if the packet is destined for the local node or if the next forwarding hop would compress out the checksum again.)

  Eliding the UDP checksum and recomputing it during decompression destroys the end-to-end nature of the UDP checksum, which is, among other purposes, useful for detecting incorrect decompression. Therefore, checksum elision is only advisable if there is another mechanism that will check the reconstituted packet, including the addresses (as the UDP checksum would do using the pseudo-header).

  As checksum elision intervenes in the end-to-end protection, the specification explicitly forbids its use unless authorized by the source of the packet, assuming that the source knows best how big the damage from undetected corruption would be. (Unfortunately, the damage might be done at an unintended receiver if the corruption implicates the destination address.)

  In the host that initially originates the packet, such an authorization can be relayed locally from the upper layer transport or application protocol instance. In a LoWPAN forwarding node, this authorization can be derived from the incoming LoWPAN packet, if it had the C bit set (this also enables the optimization mentioned above).

For an edge router injecting the packet into the LoWPAN, it will in general not be possible to derive the authorization. One way to view UDP checksum compression for packets originating from the LoWPAN and then forwarded to the larger Internet is that the originating node delegates the generation of the checksum that is required by the IPv6 world outside the LoWPAN to the edge router.

- The P field controls the compression of the source and destination fields of the UDP header (see Figure 2.18). For P = 00, the fields are sent in-line in full. Similar to the approach taken by HC1/HC2, the ports can be truncated (with the most-significant bits being replaced from the port number 61616 = 0xF0B0); however, by using two bits for the control field, compression for the two port numbers can be independently controlled, and space that would have been wasted for byte alignment is instead used for extending the encodable port number range.

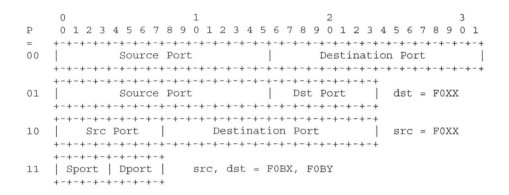

Figure 2.18 LOWPAN_NHC port number compression.

In the best case discussed above for stateless header compression (with link-local communication, UDP, 0xF0Bn ports numbers, and a Hop Limit e.g. of 1), LOWPAN_IPHC can compress the IPv6 header down to two bytes (the dispatch byte and the LOWPAN_IPHC encoding) and the UDP header to four bytes (LOWPAN_NHC, one byte for compressed port numbers, and two bytes for the checksum), see Figure 2.19; this is one byte less than the original stateless header compression. (Another two bytes can be saved if appropriate protection from higher layers allows elision of the UDP checksum.)

A more likely case in a Route-Over LoWPAN would use globally routable addresses based on a 112-bit default context and a Hop Limit that changes while routing over multiple IP hops. In the best case, LOWPAN_IPHC can compress this IPv6 header down to seven bytes (two bytes dispatch/LOWPAN_IPHC, one byte Hop Limit, two bytes each for source and destination address, see Figure 2.20). In some cases, the interface ID parts of the address will require more bytes and a CID byte might become necessary, increasing the total compressed header size including UDP to around 20 bytes. Only when the context is not at all helpful for compressing the addresses or when special IPv6 features such as flow labels are used will the compressed header size go much further beyond that.

```
 0                   1                   2                   3
 0 1 2 3 4 5 6 7 8 9 0 1 2 3 4 5 6 7 8 9 0 1 2 3 4 5 6 7 8 9 0 1
+-+-+-+-+-+-+-+-+-+-+-+-+-+-+-+-+-+-+-+-+-+-+-+-+-+-+-+-+-+-+-+-+
|0|1|1|1 1|1|0 1|0|0|1 1|0|0|1 1|1 1 1 1 0 0 1 1| Sport | Dport |
+-+-+-+-+-+-+-+-+-+-+-+-+-+-+-+-+-+-+-+-+-+-+-+-+-+-+-+-+-+-+-+-+
|          UDP Checksum         |  UDP Data...
+-+-+-+-+-+-+-+-+-+-+-+-+-+-+-+-+-+-+-+-+-+-+-+-+-+-+-+-+-+-+-+-+
       TF  N  HLM  C  S|SAM|M|D|DAM    (UDP)    C   P
       \_    IPHC base header   _/ \_NHC basehdr_/
```

Figure 2.19 Best-case LOWPAN_IPHC IPv6 packet.

```
 0                   1                   2                   3
 0 1 2 3 4 5 6 7 8 9 0 1 2 3 4 5 6 7 8 9 0 1 2 3 4 5 6 7 8 9 0 1
+-+-+-+-+-+-+-+-+-+-+-+-+-+-+-+-+-+-+-+-+-+-+-+-+-+-+-+-+-+-+-+-+
|0|1|1|1 1|1|0 0|0|1|1 0|0|1|1 0| Hop Limit     |     SA1       |
+-+-+-+-+-+-+-+-+-+-+-+-+-+-+-+-+-+-+-+-+-+-+-+-+-+-+-+-+-+-+-+-+
|     SA2       |         DA1/2          |1 1 1 1 0 0 1 1|
+-+-+-+-+-+-+-+-+-+-+-+-+-+-+-+-+-+-+-+-+-+-+-+-+-+-+-+-+-+-+-+-+
| Sport | Dport |       UDP Checksum     |  UDP Data...
+-+-+-+-+-+-+-+-+-+-+-+-+-+-+-+-+-+-+-+-+-+-+-+-+-+-+-+-+-+-+-+-+
       TF  N  HLM  C  S  SAM  M  D  DAM              (UDP)    C   P
       \_    IPHC base header   _/                  \_NHC basehdr_/
```

Figure 2.20 Globally routable best-case LOWPAN_IPHC IPv6 packet.

2.7 Fragmentation and Reassembly

IP packets come in many different sizes. The smallest (not very useful) IPv6 packet would contain just the 40-byte IP header, making its total size 40 bytes and its payload length 0 bytes. The largest IPv6 packet would be defined using the jumbogram option [RFC2675], allowing up to $4,294,967,295\,(2^{32}-1)$ bytes of payload; without this rarely used mechanism, an IPv6 packet can contain a payload of up to $65,535\,(2^{16}-1)$ bytes and may thus be up to 65,575 bytes in total size. Few subnetworks can transport packets that big efficiently; most define a *maximum transmission unit* (MTU) well below that, e.g. 1500 for (standard) Ethernet. In this section, we will discuss MTU-related issues with respect to 6LoWPAN and IPv6, but also with respect to IPv4 as this serves to illustrate some of the mechanisms and design decisions of 6LoWPAN.

On a host with a single interface, there is usually a way to find out the MTU of that interface, which allows an application to adjust the size of the packets it sends. With multiple interfaces, this choice becomes less clear. Worse, the MTU is likely to change from hop to hop on the way to the packet's destination, which makes it harder for an application to choose a size that can be transported on all the links on the path. IPv4 made only a very loose requirement on a link's MTU: an IPv4 subnetwork must be able to transmit packets of at least 68 bytes. Since most links can handle larger packets, it is not reasonable to restrict applications to sending packets of this size or smaller. Instead, IPv4 requires each node (originating host or router) to be able to *fragment* packets into several smaller ones, and each final destination must be able to *reassemble* these fragments. To distinguish the whole (unfragmented) packet from its fragments, the former is often called the *datagram*.

Note that the very small minimum MTU of IPv4 is often confused with the larger minimum *reassembly* buffer size: every IPv4 destination *must be able to receive a datagram of 576 [bytes] either in one piece or in fragments to be reassembled* [RFC0791]. Some protocols such as the original DNS (up to the introduction of EDNS0 [RFC2671]) limit themselves to never send larger datagrams. TCP can negotiate a larger datagram size via its MSS (maximum segment size) option during connection setup; implementation of this option was made mandatory to widely enable this rather important performance optimization.

Using a larger datagram size at the source actually only increases overall TCP performance if the datagram is not fragmented later in transit. The hosts engaging in MSS negotiation know about the MTU on their own links, so they can make sure this does not happen on the first and on the last hop; a mechanism for detection of the minimum MTU on the intervening links is called *path MTU detection* (PMTUD, [RFC1191] for IPv4). The algorithm employs the *don't fragment* (DF) flag of IPv4, which causes a node that would have to fragment the packet on the way forward to instead send back an ICMP *destination unreachable* error (*fragmentation needed and DF set*, often just called *packet too big*). As the path between the TCP instances can change, the algorithm needs to be adaptive: when routing changes to a new path with a smaller path MTU, the now-too-large packets will cause an ICMP packet to be generated quickly, allowing the sender to reduce its datagram size. An increase of the available path MTU can only be detected by the sender occasionally probing the path using a packet larger than the current known path MTU.

The result of the wide deployment of PMTUD is that, in samples that are likely to be representative of typical Internet traffic, less than 1 percent of packets are fragmented [Shan02]. Still, every IPv4 packet sets aside 32 bits of header space (see Figure 2.21) to enable fragmentation further on the path; the header space is reserved even if the DF bit is set.

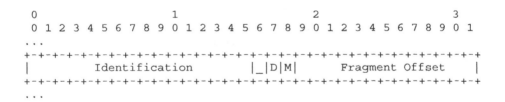

```
 0                   1                   2                   3
 0 1 2 3 4 5 6 7 8 9 0 1 2 3 4 5 6 7 8 9 0 1 2 3 4 5 6 7 8 9 0 1
 . . .
 +-+-+-+-+-+-+-+-+-+-+-+-+-+-+-+-+-+-+-+-+-+-+-+-+-+-+-+-+-+-+-+-+
 |          Identification         |_|D|M|       Fragment Offset        |
 +-+-+-+-+-+-+-+-+-+-+-+-+-+-+-+-+-+-+-+-+-+-+-+-+-+-+-+-+-+-+-+-+
 . . .
```

Figure 2.21 Fragmentation fields in the IPv4 Header (D = DF, M = MF).

IPv4 reassembly operates at the final destination of the datagram. Fragments that have the same source address, destination address, IP protocol field value, and identification field value (the *four-tuple*) are grouped together. The first fragment of a datagram is identified by its *fragment offset* being zero, the last fragment has a *more fragments* flag value of zero. (If both are zero, the datagram is unfragmented and no reassembly is necessary.) The Internet does not guarantee in-sequence delivery of packets, so the fragments of a packet might arrive out of order; fragments are positioned into the reassembly buffer based on their fragment offset (which counts in units of eight bytes and therefore only needs 13 bits instead of 16). The datagram is complete when a last fragment is present and all byte positions up to the last fragment have also been filled in by other fragments. Before the last fragment has been seen, the IPv4 reassembly process does not know how big the datagram will become.

Packets (and thus fragments of a datagram) can be lost, possibly leaving the reassembly process with incomplete datagrams. It therefore needs to employ timers to give up on reassembly of a datagram after some time. Note that in such a case the fragments that did arrive have been transmitted in vain; IP provides no way for a higher layer protocol to make use of the information, which is in stark contrast to how e.g. TCP SACK (selective acknowledgments) can use any segment that has arrived. Assuming uncorrelated losses, fragmenting a datagram into n fragments exacerbates a packet loss probability of p into a datagram loss probability of $1 - (1 - p)^n$, so a 10 percent packet loss becomes a 47 percent datagram loss if six fragments are involved. In addition, all those partially reassembled datagrams consume significant buffer space until they can finally be discarded. Outside of networks that are essentially loss-free (such as wired LANs), the use of IP fragmentation is therefore generally minimized.

To simplify the base IP header, IPv6 did away with IPv4's fragmentation fields. If a source of a datagram does want to use fragmentation, it needs to insert a separate fragmentation header as an extension header. This is the only place that fragmentation can occur: IPv6 do not provide for in-transit fragmentation. To make this more palatable, the minimum MTU was increased considerably: all IPv6 subnetworks have to provide a minimum MTU of 1280 bytes. The recommended MTU is actually 1500 bytes [RFC2460, section 5] (to coincide with the MTU exhibited by Ethernet, which is dominant on Internet paths), but the smaller minimum of 1280 bytes was chosen to leave space for tunneling headers that may have to be wrapped around the packet before transmission on an Ethernet, while also leaving some space for the IP header and further header stacks in front of a likely payload size of at least 1024 bytes. (As with IPv4, IPv6 defines a different minimum *reassembly* size: this has been set at 1500 bytes.)

Higher layers that want to send datagrams larger than the path MTU can fragment at the source using the IPv6 fragment header (Figure 2.22). Apart from the *next header* field needed to link it into the IPv6 extension header chain, its contents is quite similar to the IPv4 fields it replaces. However, the identification field has been made twice as large, as the 16 bits provided by IPv4 are no longer adequate at today's higher link speeds [RFC4963]. There also is no need for a DF flag: IPv6 packets may never be fragmented in transit, so that flag is implicitly always set.

The IPv6 specification advocates that IPv6 nodes implement PMTUD (which is defined in [RFC1981] for IPv6), so they can discover and take advantage of path MTUs larger than 1280 bytes. Given that most Internet paths have path MTUs related to Ethernet's less than 20 percent larger value of 1500 bytes, the specification reasonably also notes that resource-constrained IPv6 implementations can simply limit themselves to sending packets no larger than 1280 bytes, relieving them of the need to implement PMTUD.

In summary, while IPv6 does provide its own fragmentation for datagrams larger than the minimum MTU of 1280 bytes or the path MTU ascertained, it relies on the subnetwork layer to be able to transport packets of at least 1280 bytes. For links unable to transport 1280 byte packets natively, this means some work has to be done in their IPv6 adaptation

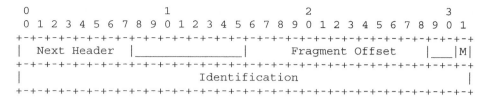

Figure 2.22 IPv6 fragment header (M = MF).

layers, this time splitting up packets into link-layer frames and reassembling them at the receiving IPv6 node.

One such link that was prominent in the 1990s was the cell-based *asynchronous transfer mode* (ATM), which in its adaptation layer type 5 (AAL5) defined a *segmentation and reassembly* (SAR) scheme in two sublayers. The lower sublayer forms cell chains using a header bit in each cell to indicate whether further cells were part of the same packet (akin to the MF bit mentioned above), and the upper layer manages padding to indicate how much space is wasted in the last cell (and adds some error checking not provided by ATM itself).

2.7.1   The fragmentation format

Compared with ATM, 6LoWPAN poses quite different requirements on its fragmentation and reassembly mechanism. On the one hand, its link layer already provides error checking as well as variable length frames, so no equivalent of AAL5's upper layer is needed. On the other hand, 6LoWPAN links, in particular with the possibility of frames being reordered during Mesh-Under forwarding, do not provide the in-sequence virtual circuit semantics of ATM, so more information than a single bit is required to splice together the right frames into one IPv6 packet.

As the requirements are quite similar to those of IPv4 fragmentation, 6LoWPAN adopts a similar mechanism, but differs in details for additional efficiency in encoding and implementation.

Instead of providing a more-fragments flag, 6LoWPAN copies the size of the packet to be reassembled (IPv6 header + IPv6 payload) into every fragment. This enables the receiving end to allocate a buffer for the whole reassembly unit upon reception of the first fragment, independent of which of the fragments actually arrives first. RFC 4944 somewhat misleadingly calls the reassembled packets *datagrams* and therefore calls the size field *datagram_size*. It is 11 bits in length, which would allow reassembled units of up to 2047 bytes (fitting nicely the actual IP-layer MTU of 6LoWPAN interfaces, which is defined at 1280 bytes).

Similar to the IPv4 Identification field, a 16-bit *datagram_tag*, combined with the sender's link layer address, the destination's link layer address and the *datagram_size*, is used to distinguish the different packets to be reassembled. Note that the tag size of 16 bits was considered sufficient for the limited link speeds used with 6LoWPAN: let's assume that the total size of the sequence of L2 packets sent for one fragmented L3 packet is at least 128 bytes (otherwise adaptation-layer fragmentation would not kick in). Even if a single source completely saturates a 250 kbit/s IEEE 802.15.4 link, it takes more than four minutes for the 16-bit tag to wrap around.

An 8-bit *datagram_offset* indicates the position of the fragment in the reassembled IPv6 packet; as in the IP layer, this counts in 8-byte units, so eight bits can cover the entire 2047 bytes. As with all 6LoWPAN frames, the first byte (dispatch byte) of a fragment indicates the type of frame; eight of the possible values for that byte (11100nnn) have been allocated for fragments with *datagram_size* spilling over into the dispatch byte so that no padding is required to accommodate the other bits. The resulting format, shown in Figure 2.23, is used for all but the initial fragment of a 6LoWPAN fragment sequence.

```
 0                   1                   2                   3
 0 1 2 3 4 5 6 7 8 9 0 1 2 3 4 5 6 7 8 9 0 1 2 3 4 5 6 7 8 9 0 1
+-+-+-+-+-+-+-+-+-+-+-+-+-+-+-+-+-+-+-+-+-+-+-+-+-+-+-+-+-+-+-+-+
|1 1 1 0 0|    datagram_size    |         datagram_tag          |
+-+-+-+-+-+-+-+-+-+-+-+-+-+-+-+-+-+-+-+-+-+-+-+-+-+-+-+-+-+-+-+-+
|datagram_offset|
+-+-+-+-+-+-+-+-+-+
 \_ dispatch _/
```

Figure 2.23 Non-initial 6LoWPAN fragment.

Assuming that most packets sent in a LoWPAN will be relatively small even if fragmented, a significant part of the fragments will be initial fragments with a fragment offset of all zeros. An optimization allows eliding that number; another eight possible dispatch values (11000nnn) were consumed for an alternative fragment format that implies a *datagram_offset* of zero (Figure 2.24).

```
 0                   1                   2                   3
 0 1 2 3 4 5 6 7 8 9 0 1 2 3 4 5 6 7 8 9 0 1 2 3 4 5 6 7 8 9 0 1
+-+-+-+-+-+-+-+-+-+-+-+-+-+-+-+-+-+-+-+-+-+-+-+-+-+-+-+-+-+-+-+-+
|1 1 0 0 0|    datagram_size    |         datagram_tag          |
+-+-+-+-+-+-+-+-+-+-+-+-+-+-+-+-+-+-+-+-+-+-+-+-+-+-+-+-+-+-+-+-+
 \_ dispatch _/
```

Figure 2.24 Initial 6LoWPAN fragment.

A node that needs to send a 6LoWPAN PDU that is too big to fit into a link-layer frame might use the following procedure:

• Set variable *packet_size* to the size of the IPv6 packet (header and payload), and *header_size* to the size of the 6LoWPAN headers that need to be in the first fragment only, such as the dispatch byte and uncompressed or compressed IPv6 headers (including non-compressed fields in the latter case). If part of the IPv6 packet header or payload (such as the UDP header) is compressed away into the compressed header, adjust *header_size* down by that amount (which will usually make *header_size* negative!). In summary, *header_size* + *packet_size* is the size of the 6LoWPAN PDU that would result if it could be sent unfragmented.

- Set variable *max_frame* to the space left in the link-layer frame after accounting for PHY, MAC, address and security headers and trailers, as well as any 6LoWPAN headers that may need to be prepended to each fragment or full packet (such as mesh headers for Mesh-Under). Note that *max_frame* may depend on the actual next-hop destination and the security and address size settings applicable for that. (Unless *header_size* + *packet_size* > *max_frame*, no fragmentation is needed and the PDU can simply be packaged into a link-layer frame.)

- Increment a global *datagram_tag* variable to a new value that will be used in all fragments of this PDU.

- Now the tricky part is to send just so much data that the first fragment is nicely filled but ends on a multiple of 8 bytes *within the IPv6 packet*. Set variable *max_frag_initial* to $\lfloor (max\_frame - 4 - header\_size)/8 \rfloor * 8$ (leaving four bytes of space for the initial fragment header).

- Send the first *max_frag_initial* + *header_size* bytes of the 6LoWPAN PDU in an initial fragment, prepending the four-byte initial fragment header to those bytes.

- Set variable *position* to *max_frag_initial*.

- Set variable *max_frag* to $\lfloor (max\_frame - 5)/8 \rfloor * 8$ (leaving five bytes of space for each non-initial fragment header).

- As long as *packet_size* − *position* > *max_frame* − 5:

    - Send the next *max_frag* bytes in a non-initial fragment, prepending the five-byte non-initial fragment header to those bytes, filling in the value of *datagram_offset* from *position*/8.

    - Increment *position* by *max_frag*.

- Send the remaining bytes in a non-initial fragment, filling in the *datagram_offset* from *position*/8.

The somewhat surprising boundary conditions of this procedure are caused by the constraint that the *datagram_offset* can only describe positions within the packet that are multiples of eight bytes; all but the final fragment may therefore not necessarily fill the space available.

Fragment reception and reassembly might operate by this procedure:

- Build a *four-tuple* consisting of:

    - the source address,

    - the destination address,

    - the *datagram_size*, and

    - the *datagram_tag*.

- If no reassembly buffer has been created for this four-tuple, create one, using the *datagram_size* as the buffer size, and initialize as empty a corresponding list of fragments received.

- For the initial fragment:

  - set variable *datagram_offset* to zero;
  - discard the four-byte fragment header;
  - perform any decoding and decompression on the contained dispatch byte and any compressed headers, as if this were a full packet, but using the full *datagram_size* for the reconstruction of length fields such as the IPv6 payload length and the UDP length;
  - set temporary variables *data* to the contents and *frag_size* to the size of the resulting decompressed packet.

- For non-initial fragments:

  - set variable *datagram_offset* to the value of the field from the packet;
  - discard the five-byte fragment header;
  - set temporary variables *data* to the contents and *frag_size* to the size of the data portion of the fragment received, i.e., minus the size of the five-byte header.

- Set variable *byte_offset* to $datagram\_offset * 8$.

- Check that *frag_size* either:

  - is a multiple of 8 (allowing additional fragments to line up with the end), or
  - $byte\_offset + frag\_size = datagram\_size$ (i.e., this is the final fragment);

  if neither is true, fail (there would be no way to fill in the remaining bytes).

- If any of the entries in the list of fragments received before overlaps the interval [$byte\_offset, byte\_offset + frag\_size$):

  - If the overlapping entry is identical, discard the current fragment as a duplicate.
  - If not, fail.

- Otherwise, add the interval to the list.

- Copy the contents of *data* to the buffer positions starting from *byte_offset*.

- If the list of intervals now covers the whole span, the reassembly is complete, and the buffer contains a reassembled IPv6 packet of size *datagram_size*. Perform any final processing that requires the whole packet such as reconstructing a compressed-away UDP checksum.

In case of failure by overlap, discard the reassembly buffer, possibly retrying the reassembly with a fresh reassembly buffer. The procedure becomes much simpler if the requirement of the 6LoWPAN format specification to perform this overlap detection is ignored.

2.7.2   Avoiding the fragmentation performance penalty

Fragmentation is undesirable for a number of reasons, many of which were listed in the 1987 paper *Fragmentation Considered Harmful* [Kent87]. The problem most often discussed is the decoupling between unit of loss (the fragment) and unit of retransmission (the entire packet), with the related inefficiencies. Possibly even more important in a resource-constrained embedded environment, the uncertainty of when the remaining fragments for a reassembly buffer will be received makes management of the resources assigned to reassembly buffers very difficult. This is probably less of an issue for more resource-heavy edge routers, but can make reception of fragmented packets by battery-operated systems with limited RAM quite unreliable. As a rule of thumb, fragmentation is marginally acceptable for packets originating from such devices (where the main problem is the reduced probability of the whole reassembled packet arriving intact), but should be avoided for packets being sent to them.

Unfortunately, the actual value of *max_frame* (the space left in the link-layer frame for IP payloads, see Section 2.7 above) is not easy to find out for a correspondent node on the Internet, as the specific layer-2 address length and security settings may not be known. Worse, it is very hard to predict from the outside how successful the header compression mechanism will be, and how much of the IPv6 and UDP headers will have to fit into *max_frame* as non-compressed fields. 6LoWPAN currently does not provide a way to perform the layer-2 equivalent of PMTUD, except by possibly correlating loss rates to packet sizes.

This situation is compounded by the fact that consecutive layer-2 hops might have different values for *max_frame*. In certain cases, the first hop in a LoWPAN might split up a packet into e.g. 80-byte fragments, only to have the next hop split each of these into one 72-byte and one 8-byte fragment! Where link-layer (mesh) forwarding is envisaged, all nodes that have to fragment packets should therefore have a general idea of the values for *max_frame* on other nodes, or fragment down to very conservative sizes.

Back to the application view: what are the packet sizes that an application can choose to make it likely not to cause layer-2 fragmentation? As a rule of thumb, of the 127 bytes available in an IEEE 802.15.4 MAC-layer frame, five will be taken for headers and trailers, and 6–18 bytes will usually be in use for addressing. Up to another 30 bytes may be taken for security, if the highest level of security is desired, but a size around 20 is more likely. This leaves around 90 bytes for *max_frame*. The new HC might compress the IPv6 and UDP headers down to 11 bytes, but might as well take 20 or more (the uncompressed size will be 49 bytes!). In summary, a UDP-based application will have to make do with payloads of 50–60 bytes or less to have a reasonable expectation that no fragmentation occurs in the LoWPAN.

Some applications such as firmware downloads may not be so frugal in their packet size requirements. If more elaborate protocols such as TCP with its MSS negotiation cannot be employed, it may be worthwhile to improve the packet arrival rate using fragment acknowledgment and retransmissions across multihop mesh paths. One such proposal is currently being discussed by the 6LoWPAN WG as a possible extension [ID-thubert-sfr].

2.8   Multicast

While the basic model of IP communication was fixed about three decades ago, an important addition that occurred in the following decade was the IP multicast model [RFC1112]. In the definition of IPv6, multicasting was considered to be a basic ingredient and was made an integral part of the specification, even replacing the subnet broadcasting mechanism defined by IPv4. The design of IPv6 and its subprotocols, in particular Neighbor Discovery (ND [RFC4861]), was certainly shaped by the powerful and quite reliable multicasting capabilities of Ethernet links.

In a multihop wireless network with unreliable links, providing an efficient multicasting capability is difficult. IEEE 802.15.4 does not itself define a multicast capability the way Ethernet does; however, a packet can be sent to a broadcast address that reaches all nodes in radio range.

To enable efficient IPv6 Neighbor Discovery without relying on multicast, 6LoWPAN provides ND optimizations, as foreseen in the ND specification [RFC4861, Section 1] – see Section 3.2. As it is not very likely that existing IPv6 applications will be ported unchanged to (or even be meaningful on) simple embedded systems, there is also limited drive from the application side that would insist on multicast capability in LoWPANs.

6LoWPAN does support Mesh-Under routing protocols that provide multicasting capability. One simple, but rather inefficient way to provide multicasting in a multihop wireless network is *flooding*: a node that wants to emit a multicast just sends it using the radio broadcast provided by IEEE 802.15.4; nodes that receive such a broadcast simply echo the multicast unless they have seen (and echoed) it before. To enable this check without a node having to remember the entire packet, 6LoWPAN defines a basic broadcast header LOWPAN_BC0 as a place to transport a sequence number – see Figure 2.25. More efficient multicast forwarding mechanisms can be provided in conjunction with the routing protocol (Section 4.2); if the basic broadcast header is not sufficient for such a forwarding mechanism, alternative routing headers can be defined.

```
   0                           1
   0 1 2 3 4 5 6 7 8 9 0 1 2 3 4 5
  +-+-+-+-+-+-+-+-+-+-+-+-+-+-+-+-+
  |0|1|LOWPAN_BC0 |Sequence Number|
  +-+-+-+-+-+-+-+-+-+-+-+-+-+-+-+-+
   \_ dispatch _/
```

Figure 2.25 The LOWPAN_BC0 broadcast header.

To perform multicasting at layer 2 as in Mesh-Under routing, the IPv6 multicast address needs to be translated into a layer-2 multicast address. 6LoWPAN reserves a part of the 16-bit short address space for this purpose. Similar to the way the lower four bytes of an IPv6 multicast address are translated into a MAC-layer multicast address, 6LoWPAN uses the lowest 13 bits of the IPv6 multicast address in forming the 16-bit short address – see Figure 2.26.

If no layer-2 mesh forwarding capability is being used, multicast packets are simply sent as radio-range broadcasts, using the 16-bit short address of all ones (FFFF) as the MAC-layer destination address.

Since the rules for forming IP multicast addresses differ from those for unicast addresses, they are also best compressed in a different way. The LOWPAN_IPHC base header includes

```
0                   1
0 1 2 3 4 5 6 7 8 9 0 1 2 3 4 5
+-+-+-+-+-+-+-+-+-+-+-+-+-+-+-+-+
|1 0 0|  low 13 bits of dstaddr |
+-+-+-+-+-+-+-+-+-+-+-+-+-+-+-+-+
```

Figure 2.26 IP multicast address to 16-bit short address mapping.

Table 2.8 D/DAM values for M = 1.

D/DAM	Inline	
0 00	48	FFXX::00XX:XXXX:XXXX
0 01	32	FFXX::00XX:XXXX
0 10	16	FF0X::0XXX
0 11	8	FF02::00XX
1 00	128	inline (address fully inline, no compression)
1 01	48	FFXX::XXLL:PPPP:PPPP:PPPP:PPPP:XXXX:XXXX, where P is the prefix taken out of the context and L is its length
1 10	–	reserved
1 11	–	reserved

the M bit for selecting multicast address compression for the destination address (multicast addresses cannot occur in the source address field). Table 2.8 shows the D/DAM values defined for the case that the M bit is set. The first four entries encode increasingly likely combinations of scope and low-order bits in the multicast address with decreasing numbers of bits sent in-line. The fifth entry is redundant with the case 0 00 for M = 0 which also carries a destination address verbatim. The sixth entry is a special case for encoding unicast-prefix-based multicast addresses [RFC3306] and multicast addresses with embedded rendezvous point (RP) addresses [RFC3956]; link-scoped IPv6 multicast addresses [RFC4489] are not covered. To limit the number of cases that need to be considered in a decompressor, some bit combinations have not been used and are reserved.

3

Bootstrapping and Security

When light switches were physically inserted into the wiring that delivered electrical power to a specific light, each of them had a very physical relationship with the circuit it was controlling. There never was a need to configure, or *commission* a physical light switch – this configuration was set up implicitly by it being wired up correctly. Installation *was* commissioning. Obviously, this no longer works for wireless light switches.

A 6LoWPAN light switch that has lost its memory when it receives a new battery has to (not necessarily in this order):

- find the LoWPAN it is going to be part of;

- establish networking parameters such as the IP address prefix and its own IPv6 address;

- establish security associations with the relevant entities in the network;

- build paths out of the node to the relevant entities, maintain those paths and possibly start forwarding for other nodes in the network;

- establish application layer parameters such as who is interested in when the light switch is operated;

- establish security associations with the relevant entities at the application layer;

- start the application layer protocol, e.g. by making known the current position of the switch.

Some of this establishment of state has to be repeated dynamically over small timescales, such as the selection of the router to be used and the routing paths used for any forwarding. The selection of routing paths is done using a *routing protocol*, see 4.2, which may also assist nodes in the selection of routers. Other parameters are less dynamic. Their setup can be structured into two phases:

Commissioning. Some of the establishment of state will require human intervention; e.g., somebody has to decide and enter into some part of the system which lights will be controlled by the light switch. This is also the time when security relationships are initialized that will enable protecting the network and its devices and applications against attackers and accidents.

Bootstrapping. After commissioning, the node is set up to operate without further human intervention. However, there may still be state that needs to be acquired, both when a device initializes (power-up with fresh batteries) or when it enters a LoWPAN (see also Section 4.1).

Note that there are a range of activities involved in bringing a node into operation, and it is not always clear whether a specific activity is commissioning, bootstrapping, or part of a dynamic protocol such as routing. The human intervention that distinguishes commissioning from bootstrapping may be mediated through a management system, making the commissioning process itself fully automatic from the point of view of the device. Little has been standardized that could be used to enable interoperable commissioning of this kind, so we will discuss it in general terms in Section 3.1. However, there is one protocol in IPv6 that is clearly related to bootstrapping: Neighbor Discovery. This protocol has been optimized for 6LoWPAN, see Section 3.2. Finally, commissioning and bootstrapping have a strong relationship with security, which is therefore discussed in Section 3.3.

3.1   Commissioning

Most 6LoWPAN devices will be manufactured as generic devices and will leave the factory without any information that would enable them to function in a specific setting. The process of providing them with this information can be (part of what is) called *installation*, *configuration* or *provisioning*. As each of these terms has a specific connotation, we will use *commissioning* as an umbrella term for all these activities; we will use *installation* for the process that makes a device available in its intended usage setting, and reserve the term *configuration* for (establishing) the set of information that is instilled in a device in the process of commissioning.

In order to enable a device to bootstrap in a LoWPAN, it needs to know some basic parameters. As many readers will be familiar with the setup of IEEE 802.11 devices, we will start with an analogy to wireless LANs (WLANs). A WLAN station that is to be commissioned to work in an IEEE 802.11 network usually needs to know two items of information:

- How to find the network, or which network is to be joined. Mostly, this requires the knowledge of an *SSID* (service set identifier); sometimes, further configuration is required for selecting the variant of IEEE 802.11 in use and the frequency band (WLAN stations are usually *frequency-agile*, i.e., can detect which of the specific channels is being used in the frequency band(s) they support).

- Security information. The station needs to know which method is used for protecting the network and the keying materials needed, e.g., a pre-shared key for "personal WPA" (wireless protected access) or username/password for certain forms of "enterprise WPA".

The information needed for setting up a WLAN base station (*access point*) also includes both an SSID and the security configuration to be used for the network; however, additional information may be required such as the frequency/channel to be used and some information about the kind of backhaul such as Ethernet or a wireless distribution system (WDS).

Similarly, a LoWPAN device needs to have some way of identifying the network as well as some security information. Often, a LoWPAN device will not be configured to be frequency-agile, i.e. it will not search the channels for available networks but be statically configured to a specific channel (see Appendix B.1 for the channels defined in IEEE 802.15.4). There is no such thing as an SSID in IEEE 802.15.4; instead, a node can assume that it has found "its" network when the security parameters and keying materials match, i.e., the integrity checks of incoming packets do not fail. In addition to enabling the device to join the desired LoWPAN, parameters may need to be set for applications to find the appropriate peers.

There is a trade-off between the efficiency of setting up state once and for all at commissioning time and the flexibility enabled by dynamically discovering this state, at bootstrapping time or even later. As an example, providing the network prefix at commissioning time would reduce the need for Neighbor Discovery Router Advertisements (RAs). But prefixes can and will change; should each such event require recommissioning? As a compromise, the device might cache information it acquired through bootstrapping.

One way to perform commissioning is to set up the devices during production at the factory (if manufacturing is by special order or as part of a just-in-time process). This means that the installer just has to pick up the right device and place it in the right setting, e.g. an electric meter in the right household. Using a just-in-time production pipeline, this can be a very efficient process. The disadvantage is that any disturbance of this process, such as a defective device or damage incurred during installation requires an exception process (e.g., obtaining an installer device and falling back to time-consuming manual configuration) or possibly even waiting for new production.

An alternative is to deliver the devices in an unconfigured state and do the target-specific commissioning during delivery, or during installation at the target. This often involves the use of a special *installer device* that communicates with the device to configure it using out-of-band means that may include barcode labels, infrared or near-field communication, USB plugs, or any other kind of wired plug-in interface.

Finally, the device could be set up using in-band communication, i.e. using 6LoWPAN communication. It is hard to do commissioning this way that is both fully automatic *and* secure, as a completely generic device cannot be distinguished from another completely generic device, so how does an installer device know that it is indeed talking to the right device? There is one item of information that *must* have been set up for each device by design of IEEE 802.15.4: the EUI-64 to be used as the 64-bit MAC address. This identifier has to be unique (although there is still a need to detect when it might not be – see Section 3.2.3) and therefore can also be used as a device ID for application purposes. However, this identifier cannot be used as a security credential: it is broadcast, for all eavesdroppers to hear, in all frames that use 64-bit addresses!

If fully automatic, secure commissioning is desired, the process that initializes a device with an EUI-64 identifier might as well provide some unique keying material to bootstrap security during commissioning.

A factory-fresh device could be in a passive mode, waiting for an installer device to instill some basic configuration. It could also be in an active mode, attempting to join a network based on its pre-provisioned credentials, waiting for some installer device to configure it over the network. That network could be an operational LoWPAN (possibly some part of the network that is run with different security parameters) or it could be a two-device LoWPAN just set up ad hoc for commissioning between installer device and the device to be installed. There could be some user interface (such as a button) to move from passive mode to active mode. There could also be a way to derive security parameters from the device, such as a barcode printed with credentials or with a reference to credentials on some server; this information might require some additional information to make use of it (e.g., to decrypt it).

If commissioning is run from factory-fresh devices that are already placed into their final positions, this could be supported by building a special setup network for the initial connection – either as a local means of communication to the installer device or via the LoWPAN to a management system.

In summary, the commissioning process is dominated by the partially orthogonal, partially conflicting axes of security and usability. This is an area where vendors are likely to see potential for market differentiation. We will come back to the security aspect in Section 3.3.

3.2   Neighbor Discovery

The IPv6 Neighbor Discovery protocol [RFC4861] is a focal point in bootstrapping an IPv6 network. A node uses Neighbor Discovery (ND) to discover other nodes on the same link, to determine their link-layer addresses, to find routers, and to maintain reachability information about the paths to neighbors that the node is actively communicating with. ND can be combined with other protocols such as DHCPv6 to obtain additional node configuration information; for resource-limited nodes in a LoWPAN, such a combination of mechanisms is often more overhead than desired. For easy reference, some additional information about the ND protocol is given in Appendix A.3.

The ND specification states that it applies to all types of links, unless specified otherwise in the relevant IP-over-X specification [RFC4861, section 1]. In the same paragraph, it makes clear that, for networks that do not fully support link-layer multicast or where the neighbor relation is not transitive, alternative protocols or mechanisms will be required. As discussed in Section 2.8, LoWPAN-wide multicast is fundamentally expensive and 6LoWPAN therefore does not make its support mandatory; in a Route-Over configuration, the link is also not transitive (i.e., if a packet from A usually reaches B and a packet from B usually reaches C, this does not mean that a packet from A usually reaches C).

For 6LoWPAN, a specification with the required alternative protocols and mechanisms is at the working group draft level at the time of writing, close to completion [ID-6lowpan-nd]. In this book, we will call this optimized form of ND *6LoWPAN-ND*.

The base ND protocol divides nodes into the traditional roles of *host* and *router*, where only the router forwards IP packets that are not addressed to itself. Routers have to perform certain additional functions in ND as compared to hosts. As many nodes in LoWPANs will be limited in their capabilities, 6LoWPAN-ND introduces a third role, that of the *edge router*, which is specialized to perform some of the more complex functions of 6LoWPAN-ND, reducing the complexity of the tasks to be performed by the other routers and in particular by the hosts. The main new concept is that of a *whiteboard* maintained by the edge routers, centralizing some of the protocol state. Also, some simplifying assumptions are made that relieve the ND protocol of some of its tasks entirely. Finally, 6LoWPAN-ND can be used to disseminate the context information that enables the higher compression efficiency of context-based header compression (as discussed in Section 2.6.2).

This section explains how 6LoWPAN-ND is used to bootstrap the nodes on the LoWPAN and to maintain the LoWPAN while in use.

3.2.1 Forming addresses

In IPv4, a node that did not have an address configured for itself had to use the *dynamic host configuration protocol* (DHCP) to obtain one from a *DHCP server*. DHCP employs a four-way message exchange to select one of possibly multiple DHCP servers and to obtain a time-bounded *lease* on an address assigned by the selected server. DHCP has been ported to IPv6 as DHCPv6 [RFC3315]. However, the larger address size of IPv6 enables the use of a simpler mechanism for address configuration: *Stateless Address Autoconfiguration* (SAA) [RFC4862]. For a refresher on the operation of SAA, see Appendix A.4.

As already mentioned in Section 2.4, 6LoWPAN simplifies the IPv6 addressing model by *requiring* that the node addresses are formed from an interface ID built from a MAC layer address and a prefix. Two such addresses are needed for each 6LoWPAN interface: a link-local address, constructed from the prefix FE80::/10, and (usually) a globally routable address, constructed from the globally routable prefix of the LoWPAN.

How does a node find out that prefix? In standard ND, routers periodically send router announcements (RAs), and, if they don't want to wait for the periodical RA, nodes can solicit one using a Router Solicitation (RS) message. Both messages are usually multicast. In this specific case, this does not pose a problem even in LoWPANs: the communication occurs between host and first-hop router, so no expensive multihop forwarding of the messages is needed.

The 6LoWPAN information option (Figure 3.1) has been both simplified and extended with respect to the standard ND prefix information option: the pair of lifetimes that provides a grace period for renumbering in standard ND has been replaced by a single valid lifetime, and the "Reserved2" field has been removed (see Appendix A.3 for the format of the standard ND prefix information option).

As an addition, there is now a space where a four-bit Context ID number (CID) can be given: this makes the prefix supplied also available (under the CID number given) for context-based header compression (see Section 2.6.2). Generally, CID = 0 will be used for the common globally routable prefix of the LoWPAN. Additional 6LoWPAN information options can be used to supply prefixes for frequent correspondents or other context entries that might improve context-based header compression efficiency. If such an additional 6LoWPAN information option is not intended to supply another prefix for the LoWPAN, the A bit will be set to zero, indicating that the prefix given is not to be used for SAA. The C bit indicates that this information option indeed is meant to occupy a position in the context. The V bit will then normally also be set, but in the lifecycle of a context entry V should be left unset both for introducing and in advance of retracting the entry: C = 1 and V = 0 means the context entry is valid for decompression but should not yet, or no longer, be used for compression.

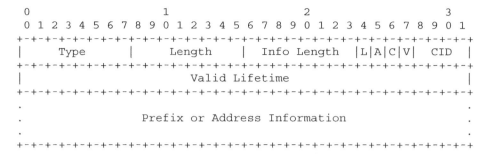

```
 0                   1                   2                   3
 0 1 2 3 4 5 6 7 8 9 0 1 2 3 4 5 6 7 8 9 0 1 2 3 4 5 6 7 8 9 0 1
+-+-+-+-+-+-+-+-+-+-+-+-+-+-+-+-+-+-+-+-+-+-+-+-+-+-+-+-+-+-+-+-+
|     Type      |    Length     |  Info Length  |L|A|C|V|  CID  |
+-+-+-+-+-+-+-+-+-+-+-+-+-+-+-+-+-+-+-+-+-+-+-+-+-+-+-+-+-+-+-+-+
|                         Valid Lifetime                        |
+-+-+-+-+-+-+-+-+-+-+-+-+-+-+-+-+-+-+-+-+-+-+-+-+-+-+-+-+-+-+-+-+
.                                                               .
.                 Prefix or Address Information                 .
.                                                               .
+-+-+-+-+-+-+-+-+-+-+-+-+-+-+-+-+-+-+-+-+-+-+-+-+-+-+-+-+-+-+-+-+
```

Figure 3.1 6LoWPAN information option.

Context-based header compression only works correctly if all nodes in the 6LoWPAN share a common view of the context. 6LoWPAN-ND therefore specifies that the complete set of context is to be *disseminated* in the entire LoWPAN, starting from the edge routers. The edge routers include the context information in their RAs, making it available to all first hop routers, which disseminate it further down the topology and so on – see Figure 3.2. The sequence number discussed below makes sure that the freshest context information wins even if another router is still sending an old version.

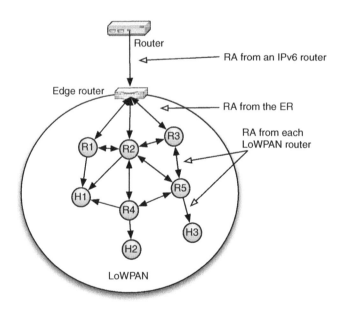

Figure 3.2 Router Advertisement dissemination.

If a large number of entries in the context are actually being used, the 6LoWPAN information options needed to send the entire context in one RA message may be up to

$16 \times (8 + 16) = 384$ bytes in size. Sending all this information in every RA would require fragmentation, compromising the delivery probability (in particular for multicast, which cannot use link-layer acknowledgment) and loading the channel. Therefore, the entire set of 6LoWPAN information options in force at one point in time is assigned a sequence number. This sequence number is included in an RA within the 6LoWPAN summary option – see Figure 3.3. (The sequence number is only valid if the accompanying V bit is set.) The unsolicited RAs that are regularly broadcast by every router only need to contain the 6LoWPAN prefix summary option. If a node notices a change in the sequence number with respect to the information it has, it can request an update by unicasting an RS message to the source of the RA; the router always responds to a unicast RS with an RA that includes the full set of prefix information.

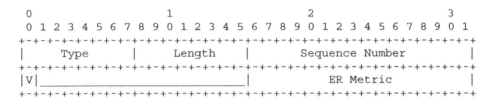

Figure 3.3 6LoWPAN summary option.

The 6LoWPAN prefix summary option also contains an *ER metric*; we will come back to that in Section 3.2.4.

3.2.2 Registration

In standard ND, the next step after forming an address would be to perform Duplicate Address Detection (DAD) on it. This is done by sending a Neighbor Solicitation (NS) to a solicited-node address, a multicast address formed as a function of the address to be validated. This process only works correctly if multicast packets are likely to reach every node that subscribes to the solicited-node address on the subnetwork, an assumption that cannot be made for 6LoWPAN.

Instead, 6LoWPAN-ND uses the edge routers as the focal point of Duplicate Address Detection. Every edge router maintains a *whiteboard* on which nodes can scribble their address and which other nodes can later read. This is done using two new ICMPv6 messages: *Node Registration* (NR) and *Node Confirmation* (NC); the entire process is accordingly called *registration*. (A detailed list of the contents of a whiteboard entry will be provided in Table 3.1 in Section 3.2.3.)

Let's start by discussing the simplest case, a node directly registering to an adjacent edge router. After obtaining a prefix for SAA as well as the address of the edge router, the host attempts to register one or more of its own addresses with the edge router by sending a Node Registration message. The edge router replies with a Node Confirmation message listing the addresses that were acceptable and includes those in its whiteboard (see Figure 3.4).

Both the Node Registration and the Node Confirmation message use the format shown in Figure 3.5. Note that the formats newly defined in 6LoWPAN-ND attempt to be similar

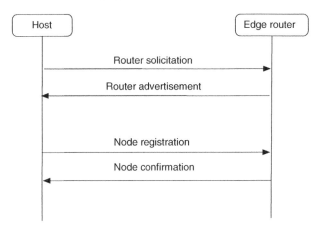

Figure 3.4 Basic router discovery and registration process with an edge router.

and compatible to the existing ND formats; this has taken precedence over any desire to
squeeze out the last couple of redundant bits (the assumption being that these messages occur
rarely and are still small enough to avoid adaptation-layer fragmentation). Type, Code and
Checksum are as usual in ND messages (we will see how Code is used in the multihop case in
Section 3.2.4). TID is a *transaction ID*, which is used in the detection of duplicate IIDs (see
Section 3.2.3). Status is used in NC messages to indicate whether the registration request
was successful. The P bit, if set, indicates that this is a primary registration; if unset, the
registration serves as a backup with additional edge routers for fault recovery. We will revisit
the TID and P fields later in this chapter.

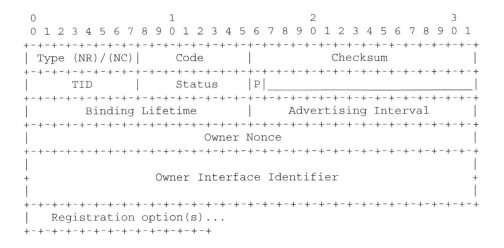

Figure 3.5 Node registration/confirmation message format.

The Binding Lifetime specifies how long the entry (entries) generated in the whiteboard by this request shall remain valid (in minutes; the 16-bit unsigned integer can indicate lifetimes up to about 6.5 weeks). The whiteboard entries, called *bindings* by analogy to a related mechanism in Mobile IPv6 [RFC3775], are *soft state*, i.e. they need to be refreshed periodically to remain in place. After the lifetime elapses, the edge router deletes the binding(s) from the whiteboard, which provides 6LoWPAN-ND's mechanism of garbage-collecting entries that have gone stale. The registering node, the *owner* of the binding, should therefore refresh the binding by sending another NR message well before the end of the lifetime, analogous to renewing a lease contract. (A second value, the Advertising Interval, describes a lifetime for the relationship of a node to its neighboring router, in multiples of 10 seconds.)

As the penultimate field of the static part of the NR/NC message, the owner of the binding is identified by its *owner interface identifier* (OII), the IID generated in the usual way (U/L bit flipped) from the EUI-64 of the node. 6LoWPAN-ND generally assumes that OIIs are globally unique (as EUI-64's should be!), but also provides a mechanism to detect the error case of multiple nodes sharing an OII (discussed in Section 3.2.3; this process is aided by the Owner Nonce).

The rest of the information in the NR/NC messages is carried in ND options, illustrated as *binding options* in Figure 3.5. Each NR message can request the registration of one or more addresses, and each NC message can confirm the registration of one or more addresses. In order to carry the addresses in one of these messages, each of them is encoded in an *address option* (see Figure 3.6).

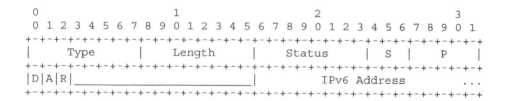

Figure 3.6 Address option format.

Type and Length are as in any ND option. The Status field indicates success (0–127) or failure (128–255) in an NC message and is unused in NR messages. There are flags that modify the request/confirmation:

D: The duplicate flag, if set in an NR, indicates that multiple registrations for one address are acceptable. This is used for registering anycast addresses.

A: The address generation flag, if set in an NR, indicates that the host is not supplying an address (the length of the included IPv6 address is 0 in this case) but that the edge router is requested to generate one. The P and S fields (see below) indicate what kind of address is requested. In an NC, the A bit is set to indicate that the address was indeed generated.

R: The removal flag, if set in an NR, indicates that this particular address is not requested for addition but for removal from the whiteboard. In an NC, a set R flag indicates that the address must not be used any longer.

Finally, P, S and the IPv6 address field compactly encode the IPv6 address to be registered/confirmed with some compression; P and S also specify the specific type of address requested in case the A-flag is set. The P field specifies the handling of the first 64 bits (the "prefix" part) of the IPv6 address, and S that of the remaining 64 bits (the "suffix"). For $P = 16$, the prefix is carried in-line in the option. $P = 17$ is used to indicate the prefix FE80::/64, any value of P between 0 and 15 for the prefix with that Context ID (CID); for $P = 16$, nothing needs to be sent for the prefix. Similarly, for $S = 0$, the suffix is carried in-line in the option. For $S = 1$, the suffix is elided and instead copied from the owner interface identifier field in the NR/NC message header, and for $S = 2$, the suffix is constructed from a 6LoWPAN 16-bit short address as defined in the 6LoWPAN format specification (or, if a non-IEEE 802.15.4 radio is in use, "as appropriate for the link layer of the LoWPAN") sent in-line. The other possible values for P and S are reserved.

The NR/NC message exchange is best demonstrated by a small example. Let's assume that a host that is on the same link with an ER wants to register two addresses to the whiteboard, both an address generated from its EUI-64 with SAA and an additional address for which the ER is asked to generate a 16-bit short address (which the host then can also use as a MAC layer address).

To register, the host sends an NR message to the link-local address of the ER. Let's say the host has just booted, therefore the Transaction ID (TID) starts with 0; a primary registration is requested, the registration is to be valid for a lifetime of 600 s (10 minutes); and the host's EUI-64, suitably modified, is used as the owner interface identifier.

Figure 3.7 shows an NR message for this registration with two address options: one for the address constructed from the Context ID 0 default prefix and the owner IID (in this case, the address is completely compressed away), and one requesting the assignment of a 16-bit short address for the construction of an address from the Context ID 0 default prefix and the newly assigned short address (in this case, the address is not sent with the NR as the A bit is set).

Figure 3.8 shows a possible NC message the ER sends in reply. The TID is echoed as 0. The status is 0, so the registration was successful. The lifetime is confirmed as 600 s. The address option for the address constructed from the Context ID 0 default prefix and the owner IID is just echoed and thus confirmed. The second address option confirms the assignment of a 16-bit short address – that address is sent back in the NC with the address option; the host then goes ahead and reconstructs the full IPv6 from the Context ID 0 default prefix and the newly assigned short address.

From now on, the host will regularly re-register before the registration lifetime expires. The re-registration differs in its TID, but also in the way the assigned short address is handled: once the host has been assigned the 16-bit short address, the host simply refreshes the binding for this address; the A bit is no longer set. Figure 3.9 shows how the second address option would look like in an NR message for re-registration (this address option is bitwise identical to an option that would attempt newly registering a 16-bit address obtained in a different way).

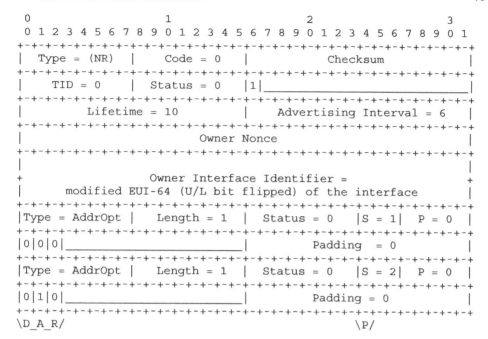

```
 0                   1                   2                   3
 0 1 2 3 4 5 6 7 8 9 0 1 2 3 4 5 6 7 8 9 0 1 2 3 4 5 6 7 8 9 0 1
+-+-+-+-+-+-+-+-+-+-+-+-+-+-+-+-+-+-+-+-+-+-+-+-+-+-+-+-+-+-+-+-+
|  Type = (NR)  |   Code = 0    |             Checksum          |
+-+-+-+-+-+-+-+-+-+-+-+-+-+-+-+-+-+-+-+-+-+-+-+-+-+-+-+-+-+-+-+-+
|   TID = 0     | Status = 0   |1|_____|
+-+-+-+-+-+-+-+-+-+-+-+-+-+-+-+-+-+-+-+-+-+-+-+-+-+-+-+-+-+-+-+-+
|          Lifetime = 10        |   Advertising Interval = 6    |
+-+-+-+-+-+-+-+-+-+-+-+-+-+-+-+-+-+-+-+-+-+-+-+-+-+-+-+-+-+-+-+-+
|                         Owner Nonce                           |
+-+-+-+-+-+-+-+-+-+-+-+-+-+-+-+-+-+-+-+-+-+-+-+-+-+-+-+-+-+-+-+-+
|                                                               |
+               Owner Interface Identifier =                   +
|          modified EUI-64 (U/L bit flipped) of the interface   |
+-+-+-+-+-+-+-+-+-+-+-+-+-+-+-+-+-+-+-+-+-+-+-+-+-+-+-+-+-+-+-+-+
|Type = AddrOpt |   Length = 1  |   Status = 0  |S = 1|  P = 0  |
+-+-+-+-+-+-+-+-+-+-+-+-+-+-+-+-+-+-+-+-+-+-+-+-+-+-+-+-+-+-+-+-+
|0|0|0|_____|    Padding  = 0           |
+-+-+-+-+-+-+-+-+-+-+-+-+-+-+-+-+-+-+-+-+-+-+-+-+-+-+-+-+-+-+-+-+
|Type = AddrOpt |   Length = 1  |   Status = 0  |S = 2|  P = 0  |
+-+-+-+-+-+-+-+-+-+-+-+-+-+-+-+-+-+-+-+-+-+-+-+-+-+-+-+-+-+-+-+-+
|0|1|0|_____|    Padding = 0            |
+-+-+-+-+-+-+-+-+-+-+-+-+-+-+-+-+-+-+-+-+-+-+-+-+-+-+-+-+-+-+-+-+
\D_A_R/                                        \P/
```

Figure 3.7 Example: Node Registration with two address options.

3.2.3 Registration collisions

The whiteboard of an edge router serves as a shared database for all nodes that have registered to that edge router, and, via the mechanisms that hold together an Extended LoWPAN, for the whole LoWPAN. As the LoWPAN is a distributed system, there is always a possibility of multiple nodes trying to create entries in that database that are in conflict with each other.

There are two levels of collision detection in 6LoWPAN-ND:

Address collision detection and resolution. If multiple nodes try to register the same IPv6 address, only one of these registrations should succeed. Each such registration is identified by the pair of the OII and the IPv6 address. A Node Registration that tries to register an IPv6 address that is already registered with a different OII at the same or another edge router is denied, giving the node an opportunity to retry with a different address. This mechanism replaces DAD within a LoWPAN and is useful mostly for ensuring the uniqueness of 16-bit short addresses and the IPv6 addresses built from them.

OII collision detection. The address collision detection and resolution mechanisms are based on the assumption that the OII is globally unique. The way that EUI-64s are allocated is supposed to ensure this in principle. However, if an error in allocating or storing EUI-64s causes two nodes to share an OII, address collision detection breaks down, leading to potentially severe malfunctioning of the LoWPAN. Such manufacturing errors have occurred with Ethernet MAC-48 identifiers, sometimes as

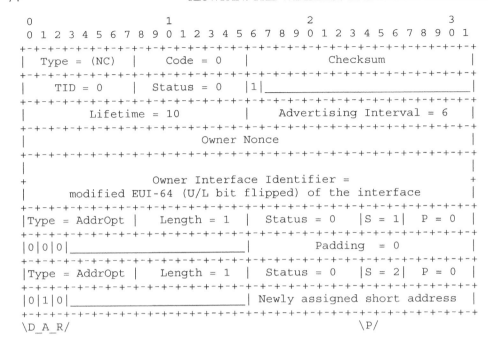

```
 0                   1                   2                   3
 0 1 2 3 4 5 6 7 8 9 0 1 2 3 4 5 6 7 8 9 0 1 2 3 4 5 6 7 8 9 0 1
+-+-+-+-+-+-+-+-+-+-+-+-+-+-+-+-+-+-+-+-+-+-+-+-+-+-+-+-+-+-+-+-+
|  Type = (NC)  |   Code = 0    |            Checksum           |
+-+-+-+-+-+-+-+-+-+-+-+-+-+-+-+-+-+-+-+-+-+-+-+-+-+-+-+-+-+-+-+-+
|    TID = 0    |  Status = 0   |1|_____|
+-+-+-+-+-+-+-+-+-+-+-+-+-+-+-+-+-+-+-+-+-+-+-+-+-+-+-+-+-+-+-+-+
|         Lifetime = 10         |   Advertising Interval = 6    |
+-+-+-+-+-+-+-+-+-+-+-+-+-+-+-+-+-+-+-+-+-+-+-+-+-+-+-+-+-+-+-+-+
|                         Owner Nonce                           |
+-+-+-+-+-+-+-+-+-+-+-+-+-+-+-+-+-+-+-+-+-+-+-+-+-+-+-+-+-+-+-+-+
|                                                               |
+              Owner Interface Identifier =                     +
|       modified EUI-64 (U/L bit flipped) of the interface      |
+-+-+-+-+-+-+-+-+-+-+-+-+-+-+-+-+-+-+-+-+-+-+-+-+-+-+-+-+-+-+-+-+
|Type = AddrOpt |   Length = 1  |   Status = 0  |S = 1| P = 0   |
+-+-+-+-+-+-+-+-+-+-+-+-+-+-+-+-+-+-+-+-+-+-+-+-+-+-+-+-+-+-+-+-+
|0|0|0|_____|         Padding   = 0         |
+-+-+-+-+-+-+-+-+-+-+-+-+-+-+-+-+-+-+-+-+-+-+-+-+-+-+-+-+-+-+-+-+
|Type = AddrOpt |   Length = 1  |   Status = 0  |S = 2| P = 0   |
+-+-+-+-+-+-+-+-+-+-+-+-+-+-+-+-+-+-+-+-+-+-+-+-+-+-+-+-+-+-+-+-+
|0|1|0|_____| Newly assigned short address  |
+-+-+-+-+-+-+-+-+-+-+-+-+-+-+-+-+-+-+-+-+-+-+-+-+-+-+-+-+-+-+-+-+
\D_A_R/                                         \P/
```

Figure 3.8 Example: Node Confirmation with two address options.

```
+-+-+-+-+-+-+-+-+-+-+-+-+-+-+-+-+-+-+-+-+-+-+-+-+-+-+-+-+-+-+-+-+
|Type = AddrOpt |   Length = 1  |   Status = 0  |S = 2| P = 0   |
+-+-+-+-+-+-+-+-+-+-+-+-+-+-+-+-+-+-+-+-+-+-+-+-+-+-+-+-+-+-+-+-+
|0|0|0|_____|Short address to be reconfirmed|
+-+-+-+-+-+-+-+-+-+-+-+-+-+-+-+-+-+-+-+-+-+-+-+-+-+-+-+-+-+-+-+-+
\D_A_R/                                         \P/
```

Figure 3.9 Example: the second address option in a refresh NR message.

a result of counterfeiting. It is therefore prudent to include a mechanism for error detection.

The assumption of uniqueness of OIIs makes the first of the two, address collision detection and resolution, relatively straightforward. When an NR message comes in, for each IPv6 address given in an address option, the whiteboard is searched for existing bindings for the IPv6 address.

- If there is no such binding, a new binding is created (potentially after checking with other edge routers – see Section 3.2.7), and a positive Node Confirmation message is returned to the node.

Table 3.1 Information content of a Node Registration binding.

IPv6 address	The IPv6 address being registered by the LoWPAN Node. This is an IPv6 unicast address of any scope.
OII	The owner interface identifier of the LoWPAN Node is used for address collision detection and resolution.
Owner Nonce	The Owner Nonce, as supplied in the last successful registration for this binding. This is used for duplicate OII detection.
TID	The Transaction Identifier of the last successful registration for this binding. This is used for duplicate OII detection.
Primary flag	Is this edge router the primary ER for the registration? Influences 6LoWPAN-ND operation between edge routers in Extended LoWPANs.
Age/lifetime	The binding age indicates how long ago the last registration message exchange took place. When the binding age reaches the registration lifetime, the whiteboard entry is discarded.

- If there is such a binding with the same OII, the binding is refreshed with the new lifetime after checking the TID to detect a possible OII collision (see below), and a positive Node Confirmation message is returned to the node.

- If there is a binding with a different OII, we have an address collision, and a negative Node Confirmation message is returned.

Table 3.1 summarizes the information kept by the edge router in a binding.

A more difficult scenario to guard against is the case where two nodes come up in the network that both believe they have the same EUI-64 and thus the same OII, too. As a general observation, it is very hard for other nodes to distinguish between these two nodes unless they add some other distinguishing property such as a random number (which, with a low probability, could also collide!). This is the reason why standard ND makes it the responsibility of each node to find other nodes it is in conflict with – only the node itself can reliably distinguish itself from others. Outsourcing the conflict detection to the whiteboard to avoid the need for multicast requires some way for the whiteboard to detect and reject requests that appear to come from a different node with the same OII as the legitimately registered node. 6LoWPAN-ND supports this with boot-time *Owner Nonces*, i.e. random numbers that are generated randomly each time a node starts up. The Owner Nonce is used to set up a registration and maintain it; requesting a conflicting registration with a different Owner Nonce points out either a duplicate OII or a node that rebooted and forgot its nonce.

In addition to the nonce, an 8-bit sequence number called the *Transaction ID* (TID) serves to correlate consecutive messages from one node. The TID is sent in each Node Registration message and echoed back by the Edge Router in the Node Confirmation message. Both TID and Owner Nonce used in the most recent Node Registration message are kept by the Edge Router as part of each binding.

The TID is not quite a normal sequence number, which would be wrapping around from 0xFF ($2^8 - 1$) to 0. Instead, after 0xFF, the number continues at 0x10 (16). This leaves the first 16 sequence numbers (0–0xF) for what we will here call *rookie bindings*, bindings of nodes that have just booted. Such a numbering scheme has been called a lollipop scheme, because it starts with a straight sequence (0–0xF) and continues in a circle from that (0x10–0xFF to 0x10 again), resembling the shape of a lollipop (see Figure 3.10).

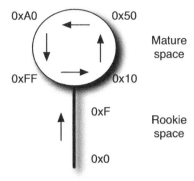

Figure 3.10 The transaction ID (TID) sequence number lollipop.

The node increments the TID after sending each Node Registration. When a positive Node Confirmation is received, if the current TID is a rookie value (less than 16), then the node sets it to 16, the first value for a mature TID. So a TID in the straight part of the lollipop denotes a node that just started/restarted and has not yet been registered.

Two TIDs i and k are compared in the following way:

- If $i < 16$ or $k < 16$, i.e., one of them is in the straight part of the lollipop, then they compare directly, i.e. a rookie value is always lower than a mature value.

- Otherwise, i.e. both i and k are in the circular part, the usual sequence number arithmetic applies, i.e. $i > k$ if $(i - k) \bmod 2^8 < (k - i) \bmod 2^8$.

A TID value is defined to be *consistent* with the preceding one if one of them could be reached from the other with between 1 and 16 increment operations, i.e. the absolute difference computed analogous to the comparison rules is 16 or less, but not zero.

A Node Registration is *consistent* with a binding if the TID in the registration is consistent with the TID in the binding, and the Owner Nonces agree. In the normal progress of initial registrations and refreshing re-registrations, the latter should be consistent with the binding.

When a Node Registration message arriving at an edge router matches the OII/IPv6-address pair of an existing binding, it is checked for consistency with the binding:

- When the new message is consistent with the binding, it appears that it is coming from the same source as the previous one and there is no OII collision. (If the TID compares as less with respect to the current TID in the binding, but the Owner Nonces agree, then the registration messages apparently were reordered and the newly arrived, but older, message should be ignored, but there is no reason to assume a collision.)

- If the TID jumps down to a rookie value, this can be interpreted as either a new node coming up competing for the existing node's OII or simply as a reboot by the node owning the registration. 6LoWPAN-ND makes the optimistic (and most likely) assumption that this is a reboot. If there is an actual collision, it will be detected when the old node attempts to refresh its registration.

- If an incoming NR neither has a rookie TID nor is consistent with the binding, then the NR is rejected as an OII collision. (With a conflicting rookie coming in to a running network, this tends to punish the incumbent, not the rookie, but the problem is at least detected.)

In summary, TIDs, enhanced by the nonce mechanism, do protect with high probability against a particular nightmare scenario: a LoWPAN that has been working well for a while develops hard-to-diagnose problems when a single duplicate OII node is coming on line.

Table 3.2 summarizes the message processing rules for Node Registration messages.

Table 3.2 Processing rules for NR messages.

Case	OII	Nonce	TID	Address	Action
Initial Registration	Unique	*	*	Unique	Accept
New Address or Movement	Duplicate	=	>	*	Accept
Duplicate Message	Duplicate	=	$\leq$	*	Ignore
Node Reboot	Duplicate	$\neq$	< 0x10	*	Accept
OII Collision	Duplicate	$\neq$	$\geq$ 0x10	*	Reject

(* = wildcard)

An actual OII collision is one of the situations that in programs are often labeled with "cannot happen". If it does, there is not necessarily a good way for the losing node to react. It could simply shut down and indicate an error through local means, waiting for human intervention and resolution of the conflict. If a high level of fault tolerance is desired and there is no good way to locally call attention to the error condition, the node could enter an emergency mode by inventing a new EUI-64 for itself using mechanisms akin to those used in the *privacy extensions for Stateless Address Autoconfiguration in IPv6* [RFC3041]. This requires stable storage and/or a good source of randomness in the node. It allows the node to communicate (if only to report the specifics of the error), but not necessarily to attain its application layer identity that may also be based on the burned-in EUI-64. In any case, an OII collision is an error caused by a faulty component and should be handled as any other fault in network management.

3.2.4 Multihop registration

The registration process becomes slightly more complicated when the node registering is not adjacent to an edge router. Nodes can also register to a LoWPAN Router in the one-hop neighborhood as long as that router indicates its ability to handle the registration by setting the M flag in its RA. The router then relays the NR to the edge router and the NC back to the node – see Figures 3.11 and 3.12.

At the time that a node sends its first NR message, its link-local address is still optimistic. Therefore, the node sends this NR with the IPv6 source address set to the unspecified address (::). The relaying router takes note of the OII in the NR and the source link-layer address (which, incidentally, should directly map into each other) and uses this state to later relay the NC response from the ER to the correct link-layer address for the node.

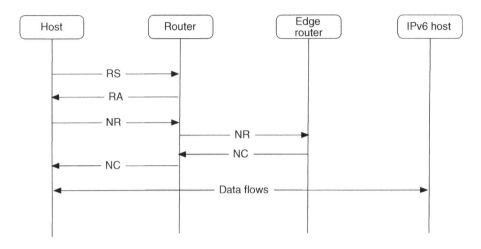

Figure 3.11 Router performing ICMP relay on the NR/NC messages.

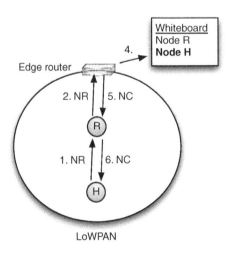

Figure 3.12 The registration process: multihop operation.

The message format for relayed NR/NC messages only differs from the one used by the hosts in that an alternative code value of 1 is used between relaying router and edge router (note that this change requires an update of the checksum of the message in the relaying router). Even though the NR/NC relaying process is a function of the router, it is not IP forwarding: the Hop Limit of the original host NR is not decremented when relaying.

Since there may be multiple edge routers, the relaying job of the router is simplified by using an anycast address for all edge routers: Each edge router configures the well-known 6LOWPAN_ER anycast address as an additional address on the LoWPAN interfaces where it serves as edge router. This means a relaying first-hop router can simply relay all NR messages to that anycast address and let routing take care of all the other hops needed in forwarding the relayed message to an ER. In other words, relaying is, if at all, only performed once at the first-hop router – from there to the edge router normal forwarding of the relayed message is performed.

In IPv4, some ICMP messages were often forged from outside a link to perform some attack such as establishing a redirect. Standard ND therefore requires most messages to be sent with the maximum Hop Limit of 255 and each receiver to check that the Hop Limit is still 255 and was not decremented by some intervening router. As a significant deviation from standard ND, the edge router needs to accept Node Registration messages that were forwarded by other routers on the way from the first-hop router to the edge router and therefore have a Hop Limit value that is lower than 255. To maintain security against outside attacks, the 6LoWPAN-ND specification requires edge routers to detect and drop Node Registration messages that do not come from a LoWPAN interface that is associated to the same LoWPAN.

Given that there may be multiple routers in the one-hop neighborhood available for registration (as is the case with the left-most host in Figure 3.2), which is the *best* that a host should choose? Obviously, if there is an actual edge router in the one-hop neighborhood, that should be chosen. Edge routers that don't have a good reason to go incognito identify themselves in their RA messages by setting the *default router preference* field ([RFC4191], the *Prf* field in Figure A.9) to high (01 binary); this field is set to medium (00 binary) for RAs from other LoWPAN routers (except for those who don't have contact with any Edge Router, which send a Preference of low, 11 binary).

But what if none of the routers in the one-hop neighborhood is an ER? This is where the *ER metric* in the 6LoWPAN prefix summary option comes in handy. The ER metric is a 16-bit unsigned integer that can be set by each router to a value indicating how good this router is as a next hop towards the outside world, expressed as some kind of routing cost metric. ERs themselves most likely set this field to 0, as they are by definition the best way out of the LoWPAN. The ER metric is intended to be used by hosts when choosing default routers; they don't need to understand the semantics of the actual metric (which is likely to derive from the routing protocol in some way – such as a simple hop count in a distance vector protocol). This leaves the hosts oblivious to the details of the routing algorithms used, as they should be. To decide between routers, hosts simply make a scalar comparison of the ER metric value, preferring routers with numerically lower ER metric values. A router with a low ER metric is not only a good default router (assuming most traffic goes outside the LoWPAN), but also is likely to be the best path to an ER that can handle Node Registrations, so the same selection should be made for registration.

In Figure 3.2, the routers in the top-level row close to the edge router (R1, R2, R3) probably can offer a better (lower) ER metric than the routers in the middle row (R4, R5). Host H1, which is adjacent to R1, R2 and R4, is therefore more likely to choose router R1 or R2 rather than router R4 as its default router and as the router to register through.

3.2.5   Node operation

LoWPAN Nodes start operation by autoconfiguring the link-local address derived from the globally unique EUI-64 of the LoWPAN interface (see Section 2.4). From the point of view of Stateless Address Autoconfiguration, the link-local address starts out as an optimistic address [RFC4429] and requires confirmation via an NR/NC message exchange with an edge router before becoming fully operational (*preferred*). Assuming that the global prefix for the LoWPAN is known at this point, the same NR/NC exchange can also be used to register the address composed of the global prefix and the same modified EUI-64. Finally, the node can go ahead and ask for assignment of a link-layer 16-bit short address from the edge router by registering an address composed of the global prefix and the IID composed from the PAN ID (usually zero) and the 16-bit short address to be assigned, as discussed in the example at the end of Section 3.2.2.

When the Node Confirmation message arrives and indicates success, the node takes note of its new 16-bit short address, adds the corresponding IPv6 address to its LoWPAN interface and changes the status of the addresses from optimistic to preferred. As the bindings established in the edge router by the registrations have a defined lifetime, the node now needs to periodically send new Node Registration messages to refresh the bindings before they run out (it can now use its link-local address as a source address as that is no longer optimistic). If a node does not receive a Node Confirmation on such a refresh message even after retries and possibly trying different first-hop routers, the addresses turn back to optimistic and the registration process needs to be restarted from the beginning.

As a result of the way LoWPAN Nodes generate their IPv6 addresses, there is a one-to-one mapping between (both 16-bit and 64-bit) link-layer addresses and corresponding link-local addresses (see Section 2.4). The same is true for the addresses generated from the global prefix. Other nodes make use of this mapping to find the link-layer addresses for intra-LoWPAN IPv6 addresses they want to sent packets to. Therefore, a LoWPAN Node never sends a Neighbor Solicitation for address resolution, which increases power efficiency and reduces implementation complexity.

But how does the node know whether a node with a LoWPAN address is in the one-hop neighborhood or needs to be addressed via a router? In a Route-Over LoWPAN, prefixes are not assumed to enable on-link decisions (the L bit is not set for prefixes announced in Router Advertisements in Route-Over LoWPANs). Instead of just the prefix, all the bits of the specific destination address are important for the on-link decision. The node may have cached information about its one-hop neighborhood from previous packets received. Or it might even optimistically attempt to transmit a packet directly to the link-layer address derived from the IPv6 destination address. The more usual node behavior is to send the packet to the default router selected via the ER metric criterion in the absence of any such cache entry.

LoWPAN Nodes do not send Neighbor Solicitations for *Neighbor Unreachability Detection* (NUD) either. Instead, the procedure recommended by the 6LoWPAN format specification applies: link-layer ACKs may be used to detect whether the packet arrived. If no acknowledgment is received after an appropriate number of retransmits, the node might delete the supposed neighbor from its one-hop neighborhood cache and retry by sending the packet to the link-layer address of its default router.

The 6LoWPAN-ND specification therefore says that the implementation of support for Neighbor Solicitation/Neighbor Advertisement messages by a node is entirely optional;

as this in turn means that a LoWPAN Node cannot rely on them being implemented by another node, the messages are indeed not very useful and it is likely that few 6LoWPAN implementations will support them at the LoWPAN host or router level.

More than NUD, routing plays an important role in getting packets to their destination in a LoWPAN. LoWPAN Routers or edge routers send ICMPv6 destination unreachable messages to indicate that delivery to a destination is not possible; nodes should therefore support processing of these messages (ICMP Type 1) [RFC4443].

In addition to unicast addresses, nodes need to support the all-nodes multicast address (FF02::1), as this is used for receiving RAs from routers. Support for other multicast addresses is not required for hosts by 6LoWPAN-ND. Multicast addresses are always considered to be on-link and are resolved as specified in the 6LoWPAN format specification (see Section 2.8).

3.2.6 Router operation

A LoWPAN Router begins operations like any other LoWPAN Node: it sets up its interface(s) and their addresses, and to perform the required Duplicate Address Detection using the whiteboard, it first needs to find an edge router or another router that has a path to the edge router.

Once the interfaces are set up, the router can start running the routing protocol that it is configured to use; once that has stabilized, it can advertise its services to other nodes using periodic Router Advertisements and by listening to Router Solicitations.

The network configuration parameters that a LoWPAN Router announces in Router Advertisements are copies of the ones it received during its own bootstrapping. As a result, these parameters originate at the edge routers. A LoWPAN Router must continue to pay attention to Router Advertisements that it receives and update its parameters whenever the sequence number in the 6LoWPAN prefix summary option increases (in the sense of the usual sequence number arithmetic). As a result, new values advertised by the edge routers flood the LoWPAN and are eventually disseminated to all routers and hosts within the LoWPAN.

Apart from its usual duties as a router, a LoWPAN Router needs to relay Node Registration messages from adjacent nodes to an edge router and relay back Node Confirmation messages to the originating node. In a Node Registration message received, the router sets the Code field to 1 to indicate that the message is being relayed, sets the IPv6 source address to its own LoWPAN address, recalculates the checksum, and sets the Hop Limit to 255. Usually, it then sends the resulting message to the 6LOWPAN_ER anycast address (a different address or set of addresses may be configured).

Nodes that are in the process of bootstrapping send their initial Node Registration messages with the IPv6 source address set to the unspecified address (::). In order to have a way to relay the Node Confirmation, the router keeps state regarding the originating node's OII to link-layer address mapping while waiting for the Node Confirmation to come back from the edge router. Once that arrives, it sets the code back to zero, sets the Hop Limit to 255, and (if the NR came from the unspecified address) uses the state to derive the link-local IPv6 destination address as well as the link-layer destination address to be put into the relayed NC. This is also a good time to cache that mapping in the router's neighbor cache.

3.2.7   Edge router operation

Most LoWPAN Nodes have exactly one interface through which they run their entire communication. The edge router is different: it also has an interface into a larger IPv6 network. As shown in Figure 1.7, this can be a simple *backhaul link* to some infrastructure unrelated to the LoWPAN, making the LoWPAN a *Simple LoWPAN*, or it can be a *backbone link* connecting to other edge routers in the same *Extended LoWPAN*. Obviously, the latter case requires significant coordination between the edge routers.

Let us start with the case of a single edge router. Again, this is a LoWPAN Router, but with additional duties:

- The edge router is the source of the network parameters disseminated in the Router Advertisements, including the LoWPAN prefix and the other context entries for the context-based header compression. Generally, edge routers will therefore require some configuration, while normal routers and hosts can bootstrap off the Router Advertisements, once commissioning parameters have been set.

- The edge router has to run the whiteboard and the two conflict detection algorithms.

- The edge router performs the routing from other IPv6 networks into and back out of the LoWPAN. To do this, it needs to run its other interfaces and possibly additional routing protocols over them. The edge router also has some protection duties in performing this forwarding: it filters out certain ICMP messages to prevent Neighbor Discovery-based attacks, and it only forwards packets into the LoWPAN for LoWPAN addresses that have been registered in the whiteboard, relieving the LoWPAN routing protocol from possibly expensively searching for a route to a node that does not exist. The system implementing the edge router function may, of course, implement other protection functions, such as a firewall or other kinds of packet filtering.

Support for Extended LoWPANs is optional in edge routers. In an Extended LoWPAN, multiple edge routers are interconnected with a backbone link, usually a higher speed, fully multicast-capable link such as an Ethernet. The backbone link has the same prefix as the LoWPAN, requiring the edge routers to bidirectionally translate between 6LoWPAN-ND on the LoWPAN interfaces and standard ND on the backbone link.

In an Extended LoWPAN, the collision detection algorithms are extended over the backbone link, using the standard ND protocol messages Neighbor Solicitation and Neighbor Advertisement. In effect, each edge router represents not only itself in the standard ND protocol, but also the collection of node addresses registered to it, much like a Mobile IPv6 home agent would for its mobile nodes' home addresses. Neighbor solicitations on the backbone link are answered by the edge routers (or other nodes on the backbone link), they never propagate into the LoWPAN. This requires that edge routers must join the solicited-node multicast addresses on the backbone link for all the registered addresses of the nodes in their LoWPANs and answer any Neighbor Solicitation with a Neighbor Advertisement that indicates the edge router's own backbone link link-layer address in the target link-layer address option.

LoWPAN Nodes moving between parts of the LoWPAN (whether by their own physical movement or by router movement or changing radio characteristics) and then sending NR refreshes to the 6LOWPAN_ER anycast address may start to reach a different edge router from the one they previously were registered with. The new edge router then simply takes over the registration, keeping the LoWPAN Node mostly oblivious to this transition. The old edge router (that had the registration previously) surrenders the registration to the one where the new NR message arrives – see Figure 3.13. Between the edge routers, on the backbone link, standard ND messages such as Neighbor Solicitation (NS) and Neighbor Advertisement (NA) are used. To enable the use of owner interface identifiers (OIIs) and transaction IDs (TIDs) in the conflict detection algorithms even over the backbone link, the NS/NA messages are extended with the *owner interface identifier option*, an option that is only used on the backbone link, never within the LoWPAN – see Figure 3.14.

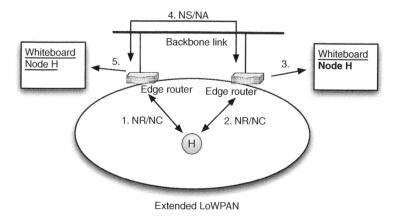

Figure 3.13 Extended LoWPAN operation as a binding moves to a new edge router.

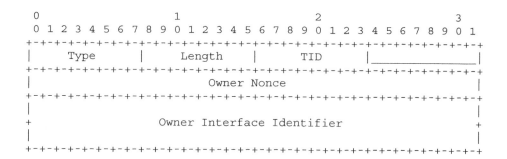

Figure 3.14 Owner interface identifier option.

3.3 Security

One important aspect of setting up a LoWPAN is establishing and maintaining its security. Wireless networks by their nature are wide open to attackers, who may overhear and inject packets being exchanged, possibly using advanced antenna technology to mount their attacks from well outside the normal range of IEEE 802.15.4 devices.

The merits of the Wireless Embedded Internet will not just be decided on technical grounds. As the proponents of consumer RFID have experienced, a few well-publicized security and privacy incidents can seriously damage the future of an emerging technology.

Another technology that took significant early damage was IEEE 802.11, which was initially designed with the assumption that weak security would be enough, under the name "wired equivalent privacy". Starting out with a compromised model is a recipe for disaster, which in this case was complemented by mistakes in implementing the model. There is a large and inventive community of security researchers (both white-hat and black-hat!) that will quickly uncover problems with any widely deployed security mechanism. Even after a quite secure version of IEEE 802.11 (IEEE 802.11i, also often referred to as "wireless protected access 2" WPA2) was established, the notion of a hard-to-protect protocol stays around as a part of the public image of IEEE 802.11, still remaining an obstacle to its deployment especially in the enterprise space.

3.3.1   Security objectives and threat models

A system cannot be called secure unless the security objectives have been defined that are being claimed to be fulfilled. These can often be grouped into three categories:

Confidentiality. Data cannot be overheard by unintended listeners, i.e., remains secret except for the authorized participants of the conversation. This is often not possible in a literal sense, but can be achieved in effect by making the information unintelligible by cryptographic *encryption*.

Integrity. Data cannot be altered by unauthorized parties, i.e. the data that is being received by an authorized participant is identical to the data that was sent by another participant. In a digital world, tampering with a message may not leave any detectable traces, so integrity is often achieved by adding cryptographic *integrity checks* to messages. Since an unaltered message from a fake sender can have the same effects as a message from the intended sender the contents of which has been tampered with, a related security objective is that of *authentication*: a message actually is originating from the source that it claims to be. A message integrity check is therefore often the same thing as message authentication.

Availability. The system is not subject to *denial of service* attacks. Note that any wireless system is subject to jamming at the radio level; however, such a jammer in a very localized network should be comparatively easy to locate and take out of service. More dangerous would be denial of service attacks from sources that cannot be easily controlled, such as the infamous "ping of death" packet that could be sent from anywhere (possibly with a fake source address) to crash certain buggy operating systems.

These security objectives may derive directly from application requirements, e.g., a factory automation system usually has very strong integrity and availability requirements, along with more or less stringent confidentiality requirements. However, security objectives can also result from the internal workings of a system: even if the data being processed in a LoWPAN are entirely public, a security element necessary for integrity that is based on a password or other secret creates an additional security objective: the confidentiality of that password.

Once the security objectives are defined, we need to understand the *threat model*: what is the attacker going to be able to do that might work against the security objectives. An important subquestion is the level of benefit that the attacker may derive from subverting a security objective. This may have a bearing on the amount of resources that the attacker can deploy. (The threat model for a bank is quite different from that for a vending machine although both are mainly attractive targets for the money they store.)

The threat model for wireless systems such as 6LoWPAN is not much different from the general threat model assumed for Internet security protocols, the *Dolev–Yao* model [Dolev81, RFC3552]: the attacker is assumed to have practically complete control over the communications channel. The attacker can read any message, remove (suppress) and change existing messages, as well as inject new messages. Again, without cryptographic support, there is no way to protect messages from reading or to detect a message that has been tampered with.

The Internet threat model assumes that end systems have *not* been compromised, mainly because it is very hard to maintain full security if that assumption cannot be made. However, the small size and distributed nature of the LoWPAN Nodes creates a rather significant threat here: in many LoWPAN deployments, it will be relatively easy to physically obtain and control at least one node in the network; the low-cost requirement will limit the level to which the nodes can be made tamper-proof. (In any case, a compromised temperature sensor can be used to inject false temperature readings by "hacking" it on the digital level or trivially by changing its temperature!) Still, measures should be taken to control the damage. In particular, it is an important additional requirement that the protection of the entire network does not depend on the integrity (and confidentiality of the memory) of each and every single node. Note that this creates hard problems, and it is therefore important to strike a balance between the potential damage and the cost of providing and maintaining the security.

3.3.2 Layer 2 mechanisms

The *end-to-end principle* [SRC81] argues that many functions can be implemented properly *only* on an end-to-end basis, such as ensuring the reliable delivery of data and the use of cryptography to provide confidentiality and message integrity. Adding a function to improve reliability on a particular link may provide some *optimization*, but can never ensure reliable delivery end-to-end. Similarly, security objectives that can only be met by protecting the conversation between two end-nodes are therefore best met by performing the cryptography at layer 3 or higher (see Section 3.3.3); there may even be security objectives that require protecting the *data* itself instead of the communication channel.

However, this does not mean that all security objectives can be met end-to-end. In particular, achieving robust availability often requires protecting the subnetwork against attackers, more so for wireless networks. Adding a first line of defense at layer 2 may also increase robustness against attacks on confidentiality and integrity. (Occasionally, there may even be a good argument that the protection against eavesdropping and forgery are already "good enough" on the wired segments of a path, suggesting that providing confidentiality and integrity just on the wireless layer 2 may be sufficient. Let's just say that this tempting argument withstands scrutiny much less often than it initially seemed attractive on a napkin.)

Not unlikely as a reaction to the early IEEE 802.11 security tribulations, IEEE 802.15.4 requires the support of rather strong cryptographic mechanisms from each node, a requirement that is fully met by most of the IEEE 802.15.4 chips available today. The encryption mechanism chosen is based on the modern algorithm AES (advanced encryption standard [AES]), which was selected by the international cryptographic community as a successor to the outdated DES (data encryption standard) in an extensive international competition between a number of strong and survivable candidate algorithms. IEEE 802.15.4 uses AES in the *counter with CBC-MAC* (CCM) mode [RFC3610], which provides not only encryption but also an integrity check mechanism.

When combining encryption with authentication, some of the authenticated information may have to be sent in the clear. AES/CCM therefore encrypts a message m and authenticates that together with (possibly empty) additional authenticated data a, using a secret key K and a nonce N. A parameter L controls the number of bytes used for counting the AES blocks in the message; m must be shorter than 2^{8L} bytes. For IEEE 802.15.4 packets, the smallest value of $L = 2$ is plenty. The nonce N is of length $15 - L$, i.e. 13 bytes for IEEE 802.15.4. The nonce is not secret, but *must* be used at most once: If an attacker has access to two messages encrypted with the same K and N, the security properties of AES/CCM are lost. The result of AES/CCM is an encrypted message of the same length as m as well as an authentication value of length M bytes, where M is a parameter that can be any even value between 4 and 16 in [RFC3610], but is restricted and extended to one of 0 (no authentication), 4, 8 or 16 bytes in IEEE 802.15.4. The authentication value can only be created correctly if K is known, so it can be checked by the receiver to ensure that m and a have not been tampered with by an unauthorized party.

AES/CCM is a quite efficient and very secure algorithm, *as long as the same nonce N never occurs twice* with the same key K. IEEE 802.15.4 builds the 13-byte nonce from the eight-byte full address of the device originating the encrypted frame (L2 packet), a four-byte frame counter, and a one-byte field occupied by the IEEE 802.15.4 security level. Appendix B.3 shows the packet subheader format involved.

So how does IEEE 802.15.4 ensure that a nonce is never used twice? The source address and security level surely repeat. So the security of the entire scheme rests on the four-byte frame counter. This allows sending up to 2^{32} encrypted frames from one source before the key that was used for these frames is "used up", assuming that the node has some stable storage that saves the current frame counter reliably even across node resets. At the current maximum IEEE 802.15.4 speed and with an assumed average frame size of, say, 32 bytes, this means that a node monopolizing the channel can do that securely with a single key for about 2^{22} seconds or about seven weeks. Even if a model based on continuous transmission results in a lower bound of key lifetime, it is clear that many applications will require rekeying within the lifetime of the system deployed.

Even if that were not the case, it is very hard to ensure that every node reliably stores the highest sequence number ever used from its EUI-64. Note that if that knowledge is lost, the key K (or the EUI-64!) has to be changed to remain secure. ISA100 (see Section 7.1) follows a slightly different approach by encoding the current TAI (international atomic time) in units of 2^{-10} seconds into four bytes of the nonce and using the last byte as a sequence number (which is unlikely to increase very much during one 2^{-10} second tick).

This makes 2^{22} seconds a hard limit of the key lifetime, meaning that new keys need to be deployed at least about every month or so. On the other hand, if reliable (and secure!) time synchronization can be achieved, this frees the nodes of the need to store the highest spent frame sequence number.

3.3.3 Layer 3 mechanisms

Even with the best link-layer security mechanisms in the LoWPAN, the data is no longer protected once it leaves the link. This makes the data vulnerable at any point that is responsible for forwarding it at the network layer, or on any link that has lesser security. Worse, an attack on the network layer might be able to divert data onto a path that contains additional forwarding nodes controlled by the attacker.

End-to-end security that protects the conversation along the entire path between two communicating nodes is therefore an important element of any robust security system, so much so, that this requirement became a banner feature in the development of IPv6. The resulting security functions have been ported back to IPv4 and are independent enough of the IP version that they are known as *IPsec* [RFC4301].

IPsec has two main components: packet formats and related specifications that define the confidentiality and integrity mechanisms for the actual data, and a key management scheme called IKE (Internet key exchange [RFC2409], updated by IKEv2 [RFC4306]). The relatively complex set of protocols that constitute IKE is generally considered a poor fit for the requirements of LoWPANs, so we will not discuss it here; see Section 3.3.4 for some considerations on alternative schemes for key management.

IPsec defines two packet formats for cryptographically protected data: the IP authentication header (AH) [RFC4302], which provides integrity protection and authentication only, and the IP encapsulating security payload (ESP) [RFC4303], which combines this function with confidentiality protection through encryption.

AH got a bad name with IPsec implementations for IPv4, as it protects not only its payload but also the addresses in the enclosing IP packet header. This detects and rejects the tampering with IP addresses performed by NATs, which are essential to the operation of IPv4 networks. Although AH is perfectly useful in the NAT-free IPv6 environments, this problem has led most IPsec implementers to focus on ESP, which we will do here, too.

Figure 3.15 shows the ESP format. In contrast to other IPv6 extension headers, which simply precede their payload, ESP encloses ("encapsulates") it, which is somewhat logical, given that the payload is encrypted. The part of the format marked "C" on the right is encrypted (confidentiality protected), the larger part marked "I" is subject to integrity protection.

The security parameters index (SPI) identifies the specific security parameters, including the keying material, used for this direction of the conversation (*security association*). For unicast packets, the SPI is an identifier of local significance to the receiver, i.e. it is assigned by the receiver in a way that facilitates its local processing of the incoming ESP packets. The sequence number is an unsigned 32-bit number that increases by one for each packet sent on the security association; it can also be the lower 32 bits of a 64-bit sequence number kept in the security association. The payload data, together with the trailer consisting of the padding, the pad length and the next header field, are the part of the packet that is encrypted; what is shown in the figure is the unencrypted/decrypted version of the data.

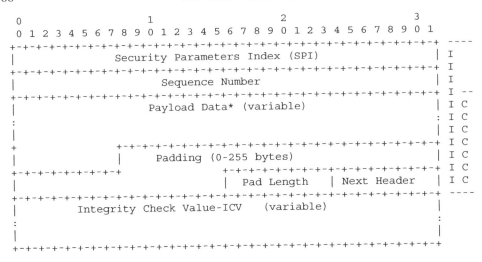

Figure 3.15 Encapsulating security payload (ESP) packet format.

The next header field specifies how the decrypted data is to be interpreted by the receiver. It can identify the payload as an IP packet (*tunnel mode*) or as some transport data such as a UDP header followed by a UDP payload (*transport mode*). Tunnel mode is most useful for security gateways ("VPN gateways"); transport mode is a more compact way to obtain end-to-end security as no second IP header needs to be sent.

The padding (the length of which is given by the eight-bit pad length field) can be added to round up the length of the payload and trailer (including pad length and next header) to a multiple of four bytes, which is the minimum required alignment of the beginning of the ICV field. (Actually, the encryption algorithm might pose more stringent alignment requirements, such as a full AES block of 16 bytes for modes such as AES-CBC [RFC3602]; this is not an issue with streaming modes such as the AES-CCM most likely used in LoWPANs.)

Finally, the integrity check value (ICV) is used with the other information in the ESP header, payload and trailer to check the integrity of the packet. The length of the ICV is defined in the security association.

LoWPAN Nodes are likely to have hardware for AES/CCM encryption, decryption and integrity check processing. This hardware is generally available in a way that allows software above the link layer to use it, making AES/CCM for ESP [RFC4309] the obvious candidate for the cryptographic suite used to achieve end-to-end confidentiality and integrity/authentication. Figure 3.16 shows how the encrypted payload and the ICV look with AES/CCM. The encrypted payload starts with an explicitly transmitted initialization vector (IV) of eight bytes. (This is prepended by three bytes of *salt* in the security association to produce an 11-byte nonce going into the CCM algorithm – the AES/CCM for ESP transform uses CCM with L = 4 to provide support for very large jumbograms.) The encrypted payload follows (the payload has been fitted with a trailer before encryption, appropriately padded to a multiple of four bytes). Finally, the ICV is appended. [RFC4309] provides for ICV sizes of eight bytes and 16 bytes as well as optionally 12 bytes; which of these values is used needs to be defined in the security association.

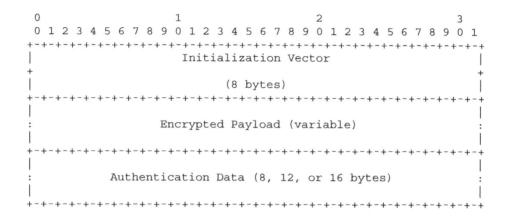

Figure 3.16 ESP payload encrypted with AES/CCM.

In summary, it can be said that ESP with AES/CCM is not necessarily too heavyweight for end-to-end encryption and integrity checking between a LoWPAN and its correspondent nodes. With most IEEE 802.15.4 capable chips, the processing overhead is as small as it possibly could be. The per-packet overhead could be a bit less with 1–4 bytes of overhead for padding (including pad length) and eight bytes for the explicitly transmitted initialization vector (IV), but it is not too bad, if solid end-to-end cryptographic protection is indeed needed. (If more efficiency is really needed, a special version of the AES/CCM transform could be defined that creates even less overhead, or preferably some special form of stateless header compression could be added to LOWPAN_NHC, e.g. by deriving the initialization vector from other information present in the MAC-layer packet.)

3.3.4 Key management

One lesson was really driven home by the IEEE 802.11 WEP disaster even before all of WEP's cryptographic security flaws surfaced: a single key shared by all devices on the network and essentially unchangeable without touching all these devices does not scale. Larger WLAN sites quickly abandoned WEP for L3-based security ("VPNs") or enhanced the L2 key management by schemes based on 802.1X and finally 802.11i.

While the username/password-based authentication schemes usually used with 802.11i "enterprise" style WLANs are not really applicable to 6LoWPAN, it is worthwhile looking at the key management aspect. At the end of the EAP-based authentication, the WLAN station and the WLAN access point share a secret, the *pairwise master key* (PMK). However, this is not used for actually encrypting traffic: the two devices then go on to derive a *pairwise transient key* (PTK) from the PMK. A new PTK is derived by the devices whenever the existing one is "spent" cryptographically.

In addition to the pairwise key useful for unicast traffic between WLAN station and WLAN access point, each WLAN access point creates a random *group transient key* (GTK) in order to be able to send broadcast frames to all WLAN stations associated with it. The GTK is usually changed with a period of a couple of hours and sent to each station via unicast using the established pairwise keys.

The specific key management scheme used here is not easily adaptable to 6LoWPAN as the situation in a WLAN is quite different: the WLAN infrastructure is essentially composed of WLAN access points (think edge routers), connected via Ethernet (think backbone link) which is usually outside the WLAN security considerations. WLAN stations directly associate with a WLAN access point. Also, the processing, storage, power and user interaction capabilities of WLAN nodes are usually beyond those of 6LoWPAN devices.

However, the example is useful to illustrate some key concepts of key management:

Long-term keys like the PMK are never directly used for encrypting traffic. Instead they take part in relatively infrequent cryptographic operations such as rekeying that result in short-term keys. (In 802.11i, the PMK is not really long-term as it is produced by each new authentication, except when the WLAN is in *pre-shared key* (PSK) mode.)

Short-term keys like the PTK are used to actually encrypt traffic. They can be efficiently replaced once their cryptographic power has been used up, e.g. by applying *key derivation functions* to some long-term keys and some nonces exchanged securely.

Pairwise keys are used between two entities to encrypt and authenticate. The resulting authentication is strong, as one of the entities can be pretty sure the other one sent the data if it wasn't itself.

Group keys are used for functions such as broadcasting. The resulting authentication is relatively weak, as one of the entities can never be sure whether the claimed source sent the data or some other member of the group, all of which know the key.

At the time of writing this section, one can only speculate about the ways these concepts will be used in LoWPANs. In an ISA100 context, there is a special security manager that is responsible for performing all key management. Devices are provisioned with *join keys* that are then used to obtain shorter-term keys such as a link-layer group key and transport layer pairwise keys from the security manager. However, at the link layer, the communication for the join process is "protected" only by a well-known key, opening up the protection of the network.

Let's end this section with simply repeating its main take-home message. As has been amply demonstrated by WEP, it is not a good idea to continue using a key over and over, especially with stream ciphers like WEP's RC4 or IEEE 802.15.4's AES/CCM. CCM completely loses its security if the same IV value is used more than once for a given key. [RFC4309, section 9] therefore rightly admonishes the reader not to use AES/CCM with statically configured keys: *"Extraordinary measures would be needed to prevent the reuse of a counter block value with the static key across power cycles. To be safe, implementations MUST use fresh keys with AES CCM."* A key management scheme is *required* in order to be able to properly take advantage of IEEE 802.15.4's superior security technology.

4

Mobility and Routing

This chapter considers IP mobility and routing issues related to 6LoWPAN. There are several causes of mobility in embedded systems utilizing low-power wireless technology. Many systems require support for *node mobility* of wireless devices. Often devices may be integrated into moving machines, carried by people or animals, or tagged on equipment or supplies. In other applications the edge routers themselves may be mobile, changing their *point of attachment* on the Internet, thus requiring us to deal with *network mobility*. While this kind of physical mobility of edge routers or nodes is easy to comprehend, there are also other more subtle causes of *topology change* in which nodes don't actually move. In wireless networks, the radio channel is continually changing due to fading effects, where changes in the environment can drastically affect radio connectivity between nodes. These radio changes often force nodes to make use of alternative routes and even to change LoWPANs without physically moving. Furthermore, autonomous embedded devices may run out of battery, fail or make use of long sleep cycles, all of which may cause changes in network topology. These basic causes of mobility along with core IPv6 and 6LoWPAN solutions for dealing with them, are covered first in Section 4.1.

Although IPv6 and 6LoWPAN have techniques to be able to prevent mobility within a LoWPAN from affecting network operation, to deal with address changes caused by nodes moving between LoWPANs, and to mitigate the effects of network mobility – the Internet Protocol does not calculate routes. This is left to a *routing protocol*, which maintains *routing tables* in routers. A routing table contains entries used by a router to determine the next hop to which a packet should be forwarded. In Section 4.2 we look at IP routing protocols useful within LoWPANs, such as those from mobile ad hoc network (MANET) and routing over low-power and lossy networks (ROLL) working groups, along with *border routing* between LoWPANs and the Internet. The basics of forwarding and routing mechanisms for 6LoWPAN were covered in Section 2.5.

6LoWPAN only supports IPv6, whereas the vast majority of Internet traffic is still using IPv4. Thus the issue of *IPv4 interconnectivity* is important for deploying 6LoWPAN in real networks. Although most operating systems, routers and core networks support IPv6 today, the majority of internet service providers (ISPs) and operators of Web servers are still planning for transition. In many enterprise and industrial applications, IPv4 interconnectivity may not be an issue if traffic between edge routers and servers occurs over a local area network (LAN) such as Ethernet. If communications with 6LoWPAN devices occurs over the Internet, then it will probably be necessary to deal with IPv4 interconnectivity. Luckily the IETF has designed an array of techniques for the transition from IPv4 to IPv6, allowing them to interwork smoothly. Techniques useful for integrating 6LoWPAN and IPv4 networks are discussed in Section 4.3.

4.1 Mobility

LoWPAN Nodes, for example in asset management, often tend to be mobile. In some cases, such as with body area networks, the network itself may even be mobile. Figure 4.1 shows a typical industrial asset management scenario, where a fork-lift is moving goods between warehouses and an assembly facility. A wireless embedded network in such an application may be used in several ways, including to track the fork-lift itself, the goods being moved and personnel at the plant. All of these uses of the wireless network require us to deal with node mobility between edge routers in the same LoWPAN, between LoWPANs and possibly between network domains. At the same time, active data flows may be in progress and application servers may need to know how to reach tracked devices. This section considers what causes different kinds of mobility and how IPv6 and 6LoWPAN can deal with mobility.

Figure 4.1   An industrial asset management application where mobility is common. (Reproduced by Permission of © SENSEI Consortium.)

4.1.1 Mobility types

Mobility in IP networks technically is the act of a node changing its topological point of attachment. Koodli and Perkins distinguish the following two kinds of mobility [Koodli07]:

Roaming: A process in which a mobile node moves from one network to another, typically with no existing packet streams.

Handover: A process in which a mobile node disconnects from its existing point of attachment and attaches itself to a new point of attachment. Handover may include operations at specific link layers as well as at the IP layer in order for the mobile node to be able to communicate again. One or more application packet streams typically accompany the mobile node as it undergoes handover.

Mobility can alternatively be described with the terms micro-mobility and macro-mobility. *Micro-mobility* refers to mobility that occurs within a network domain. In 6LoWPAN we can consider micro-mobility to refer to the mobility of a node within a LoWPAN where the IPv6 prefix does not change, which is the definition used here. *Macro-mobility* on the other hand refers to mobility between networks. In 6LoWPAN we consider macro-mobility to refer to mobility between LoWPANs, in which the IPv6 prefix changes. In relation to the previously defined terms, we can consider micro-mobility to require only handover, whereas macro-mobility is a process of roaming and handover. Figure 4.2 illustrates these forms of mobility with regard to 6LoWPAN. The process of node 1 moving its point of attachment from one edge router to another within the same Extended LoWPAN is clearly a case of micro-mobility. The IPv6 address of the node remains the same, and to the remote server it seems that nothing has changed. In contrast, when node 2 moves from the Extended LoWPAN to a different LoWPAN it is roaming to a network with a different IPv6 prefix, therefore the IPv6 address of the node will change as well.

Before looking at solutions for dealing with mobility, it helps to understand why mobility happens in the first place. In wireless networks there are a number of things that may cause a network to make a change in topology. The causes of topology change can be categorized simply as physical movement, radio channel changes, network performance, sleep schedules and node failure:

Physical movement: The most evident reason for mobility is when nodes in a network physically move in relation to each other, which changes the wireless connectivity between pairs of nodes. This may cause nodes to change their point of attachment.

Radio channel: Changes in the environment cause changes in radio propagation, called fading. These changes often require topology change even without physical movement, especially with simple radio technologies.

Network performance: Packet loss and delay on wireless networks may be caused by poor signal strength, collisions, overloaded channel capacity or node congestion. High packet loss may cause a node to change its point of attachment.

Sleep schedules: Especially battery powered nodes in wireless embedded networks use aggressive sleep schedules in order to save battery power. If a node finds itself attached to a sleeping router without a suitable duty cycle for the application, this may cause the node to move to a better point of attachment.

Node failure: Autonomous wireless nodes tend to be prone to failure, for example due to battery depletion. The failure of a router causes a topology change for nodes using it as their default router.

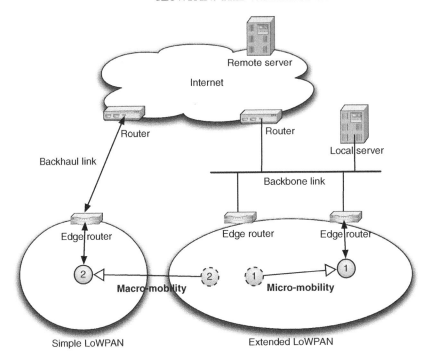

Figure 4.2 The difference between micro-mobility and macro-mobility.

The types of mobility explained above have been described in the context of what is called *node mobility*, when a single node moves between points of attachment. There is another class of mobility in which an entire network moves its point of attachment – this is called *network mobility*. An example of network mobility is shown in Figure 4.3. When considering 6LoWPAN, network mobility occurs when an edge router changes its point of attachment while nodes in the LoWPAN remain attached to it. The kind of network mobility that affects 6LoWPAN is clearly macro-mobility, when the IPv6 address of the edge router changes, as this affects the addressing of all nodes in the LoWPAN. The next section gives an overview of solutions for dealing with mobility.

4.1.2   Solutions for mobility

When a node changes its point of attachment, there are a number of things that must be handled for it to start participating in the IP network using the new point of attachment and to resume data flows again depending on the type of mobility and the solutions used for roaming and handover:

- Establish a link by performing commissioning (as in Section 3.1).

- Configure an appropriate IPv6 address by bootstrapping (as in Section 3.2).

- Deal with security and firewall settings (see Section 3.3).

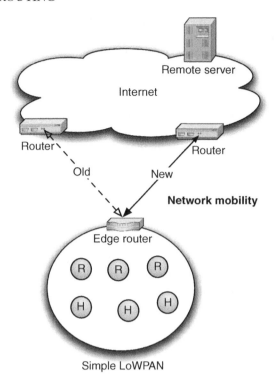

Figure 4.3 Network mobility example.

- Update any domain name space (DNS) entries to the appropriate IPv6 address (if the address changes).

- Notify the application layer and maintain any application layer identifiers or registrations (if the IPv6 address changes).

Depending on the type of mobility that occurred and the solutions used for dealing with roaming and handover, some of these steps may not be necessary, or the amount of work needed for each may differ greatly.

When micro-mobility occurs between attachment points that are part of the same link (e.g. an Ethernet backbone with multiple WiFi access points), the link layer may be able to deal with the mobility without any noticeable changes to the network layer. Common link layers that deal with mobility include cellular systems such as GPRS or UMTS which perform handovers maintaining the same IP address, and WiFi which performs Ethernet bridging. For 6LoWPAN, Mesh-Under link-layer techniques may deal with micro-mobility inside a LoWPAN, although no well-known technologies for this exist today. Low-power wireless link-layer technologies such as IEEE 802.15.4 tend to leave mobility to be dealt with by the network layer. Networks such as IEEE 802.15.4 work in a manner that all topology changes are *node-controlled* (such as in WiFi) rather than *network-controlled* (such as in cellular systems).

Neighbor Discovery for 6LoWPAN [ID-6lowpan-nd] includes a built-in feature for dealing with micro-mobility in Extended LoWPAN topologies, such as that shown in Figure 4.2. This is achieved using an *ND proxy* technique and whiteboard synchronization between edge routers, allowing a node to keep the same IPv6 address regardless of its point of attachment within the Extended LoWPAN. 6LoWPAN-ND was described in detail in Section 3.2.

Macro-mobility always involves a change of IPv6 address for a node. Such an address change can be dealt with in several ways to minimize the negative impact on the application layer. The simplest way to deal with this is for the application to simply restart when detecting a change in IP address. This is commonly used when the node acts as a client, such as with most mobile Internet hosts today, and may be applicable to simple 6LoWPAN client applications as well. If a node acts as a server, and must be reachable by any time from corresponding nodes, macro-mobility is a real challenge. One way of dealing with this in 6LoWPAN applications is at the application layer, using for example the session initiation protocol (SIP), uniform resource identifiers (URIs) or a domain name server (DNS). Mobile IPv6 [RFC3775] provides one way of dealing with this at the network layer, by maintaining a home address on behalf of mobile nodes which does not change, thus requiring no DNS changes. Mobile IPv6 is applicable to edge routers, but is considered to be too heavy a solution for Simple LoWPAN Nodes. It may be possible for edge routers to perform Mobile IPv6 on behalf of LoWPAN Nodes using the proxy Home Agent concept, which we will discuss further in Section 4.1.5.

When an edge router itself changes its point of attachment along with nodes in its LoWPAN, we are dealing with a network mobility problem. The IPv6 address of the edge router changes, and thus the IPv6 prefix of the LoWPAN changes as well. Network mobility for LoWPANs could be dealt with by using Mobile IPv6 for the edge router and all nodes in the LoWPAN (which is not actually practical for LoWPAN Nodes), or by applying the network mobility (NEMO) [RFC3963] solution presented in Section 4.1.7.

Although the changes in IPv6 address caused by roaming can be dealt with using these mobility solutions, they shouldn't be confused with routing protocols which maintain paths between nodes in IP networks. There are, however, interactions between mobility and routing. The decision to change point of attachment is often made by a routing protocol, although any change in IP address that might have caused is left to a mobility solution. We can think of a routing protocols as being able to cope with micro-mobility. Routing protocols are discussed in Section 4.2.

4.1.3 Application methods

In many cases, there may be a need to deal with the effects of mobility at the *application layer* – including IP address changes caused by roaming and service degradation caused by handover. This might be required if the LoWPAN or IP backbone in use by the application does not have mechanisms for dealing with mobility at the network layer available. In addition, dealing with mobility at the application layer may even provide better optimization for specialized applications. Issues related to the design of the application layer using 6LoWPAN are discussed in detail in Section 5.2.

When a handover does occur, one thing that needs to be considered is how the transport layer protocol reacts to a change in IP address for one of the end-points, or to a temporary disruption in service. 6LoWPAN applications often make use of UDP as a transport, which is resilient to changes in IP address as each datagram is independent. An application using UDP still needs to deal with the change in IP address, correlating the new address to the same end-point. TCP is not able to deal with IP address changes, and thus a TCP connection would be broken after an IP address changes for either end-point. The stream control transport protocol (SCTP) does have a mechanism for dealing with multiple IP addresses [RFC2960], but has the same problems for use over 6LoWPAN as TCP, which are discussed in Section 5.2.

In order for an application server communicating with mobile 6LoWPAN Nodes with changing IPv6 addresses to function, it needs to use some sort of unique and stable identifier for each node. Examples include the EUI-64 of the node's interface, a URI, a *universally unique identifier* (UUID) [RFC4122] or for example a *domain name* resolved using the DNS. Entries in the DNS by nature provide a mapping from a domain name to the IPv6 address of the corresponding 6LoWPAN Node. Care must be taken when using the DNS for such purposes as updates may not propagate immediately, although by using a client to send *dynamic updates* [RFC2136] and with careful *Time to Live* (TTL) settings it may be useful for some applications. Other identifiers need to be mapped somehow to the IPv6 address of the node, except for applications where the IPv6 address is irrelevant or the identifier is carried inside each application payload.

Some application protocols have methods built in for dealing with mobility. The session initiation protocol (SIP) [RFC3261] includes a universal way of identifying users called a SIP URI. This URI could be used for the purpose of tracking 6LoWPAN Nodes, taking a form such as node10@home.example.com. Furthermore SIP has methods for indicating that a session end-point has changed during an active session using a Re-INVITE message. The SIP header format is unfortunately too verbose for use over 6LoWPAN without modification, although some techniques to do this have been proposed. The use of SIP over 6LoWPAN is considered further in Section 5.4.6.

4.1.4 Mobile IPv6

The mobility of nodes on the Internet can be dealt with at the network layer using a protocol called *Mobile IP* (MIP), which was originally designed for IPv4 [RFC3344]. Later an updated version was developed for use with IPv6 called Mobile IPv6 (MIPv6) [RFC3775]. This new version takes advantage of IPv6 mechanisms and provides route optimization. The goal of MIP is to deal with the mobility roaming problem by allowing a host to be contacted using a well-known IP address, regardless of its location on the Internet. Mobile IP does this using the concept of a *home address*, which is associated with a host's *home network*. When a host is away from its home network, and attaches to another network domain (called the *visited network*), the new IP address it configured there is called its *care-of address*. A node communicating with a mobile node roaming in a visited network is called the *correspondent node*. The concept of 6LoWPAN is for very simple IPv6 nodes to be able to participate in IPv6 networks over low-power, low-bandwidth wireless links. Can Mobile IP be used to solve the problem of node and network mobility also in 6LoWPAN? This section introduces the way MIPv6 works, and discusses its applicability to 6LoWPAN mobility problems.

Normally forwarding in IP networks is handled only by routers, whose route tables are maintained by routing protocols. Mobile IP works using a special kind of routing functionality, which is *host controlled*. This concept is called a *binding*, and is implemented by an entity called the *Home Agent* (HA). A Home Agent must be present in the home network domain of a mobile node in order to use Mobile IP. The Home Agent is responsible for maintaining a binding between a node's permanent home address and the temporary care-of address used while roaming in a visited network. The HA then acts as a forwarder of traffic destined to and coming from the mobile node while roaming. A good analogy is that of a post office temporarily forwarding mail to a new address when someone moves.

Figure 4.4 illustrates basic MIPv6 functionality. When a mobile node roams to a visited network, it uses MIPv6 in the following way to maintain global connectivity via its home address:

- After detecting that the subnet has changed, and that the node is no longer in its home network, it sends a MIPv6 binding update message to its HA. If the node doesn't know the location of its HA or its home prefix, there are methods to discover both. The binding update is acknowledged by the HA with a binding acknowledgment. These messages must be secured using e.g. IPsec methods.

- A bidirectional IPv6-in-IPv6 tunnel is then set up between the HA and mobile node for exchanging data packets. When an incoming packet arrives from a corresponding node to the home address of a mobile node at its home network, it is intercepted by the HA using an ND proxy technique. The packet is then encapsulated in another IPv6 header with the destination address set to the mobile node's care-of address.

- After receiving and decapsulating the packet, the mobile node can then respond through the IPv6-in-IPv6 tunnel through its HA back to the corresponding node. Alternatively, the mobile node can respond directly to the corresponding node using its care-of address.

- This is a case of *triangular routing*, where a more optimal path directly between the mobile node and corresponding node is possible. MIPv6 includes *route optimization* to avoid triangular routing. After executing a *reverse routability test* and possible security association between the correspondent node and mobile node, they can begin to communicate directly. First a binding update is sent to the corresponding node. Then data traffic is exchanged using IPv6 extension headers to properly indicate the actual home address of the mobile node.

- Now communication can continue directly between the correspondent node and the mobile node.

As 6LoWPAN is an adaptation for IPv6, it would make sense to apply MIPv6 for dealing with the roaming of 6LoWPAN Nodes while still maintaining a constant IPv6 address. MIPv6 could also be used by edge routers to deal with network mobility. Unfortunately LoWPAN Nodes have tight requirements on node complexity and power consumption, and wireless links have limited bandwidth and frame size available. MIPv6 on its own has several problems that limit its usefulness for 6LoWPAN.

In order for MIPv6 to be applied to 6LoWPAN Node mobility, it would have to be implemented on LoWPAN Nodes. The use of MIPv6 as defined in [RFC3775] with 6LoWPAN has the following problems:

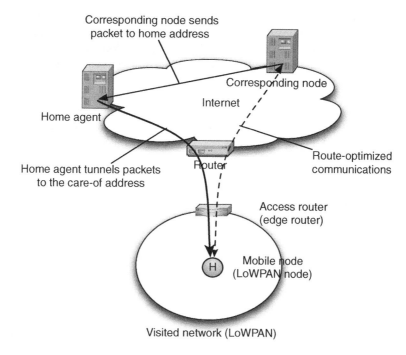

Figure 4.4 Example of Mobile IPv6 used with 6LoWPAN.

- IPv6-in-IPv6 tunneling between the HA and LoWPAN Node would incur large header overheads as the encapsulated IPv6 and transport headers cannot be compressed by the existing compression methods.

- The requirement of IPsec security associations between MIPv6 entities may be unreasonable for LoWPAN Nodes. See Section 3.3 for details on 6LoWPAN security.

- The added complexity of implementing MIPv6 in terms of code size and RAM may be unjustifiable for LoWPAN Nodes.

- In domains with large LoWPANs and frequently mobile nodes, the traffic burden caused by MIPv6 may be too much for low-bandwidth wireless links.

- Route optimization adds an even greater burden on nodes, as state must be maintained for every active correspondent node.

The application of MIPv6 directly on LoWPAN Nodes would need further study. Substantial optimization of MIPv6 would be required to run it directly on the nodes. Work has been done on a proposal for a LoWPAN adaptation technique for MIPv6 in [ID-6lowpan-mipv6] where messages are compressed and simplified. This technique would require edge routers to compress and decompress these messages to full MIPv6, and would need standardization. As an alternative solution a proxy method where the edge router or some other entity on the visited network would proxy MIPv6 functions on behalf of the LoWPAN Nodes could be feasible.

An interesting solution for this is the *proxy Home Agent* (PHA) concept, which is discussed in Section 4.1.5. The applicability of the network-based local mobility management protocol *proxy Mobile IPv6* (PMIPv6) to 6LoWPAN is discussed in Section 4.1.6.

MIPv6 could also be used by the edge router, while roaming to visited networks in order to maintain a stable IP address. Using [RFC3775] this would however only handle the IPv6 address of the edge router itself as binding entries in the HA are for specific IPv6 address. Nodes in the LoWPAN would need to implement MIPv6 as well to maintain a stable IPv6 address. This is especially problematic when trying to deal with network mobility. A better solution for network mobility of LoWPANs is available using the *network mobility* (NEMO) protocol, which is discussed further in Section 4.1.7.

4.1.5 Proxy Home Agent

A *proxy Home Agent* (PHA) is an entity which performs MIPv6 functions on behalf of a local mobile node, interacts with the actual Home Agent of the node, and handles route optimization on its behalf. This greatly simplifies the functions that a mobile node needs to perform to participate in MIPv6. In 6LoWPAN this is an especially critical optimization as seen from the previous section. A global architecture for PHA is described in [ID-global-haha], where PHA functionality is described.

The PHA is located in the visited network where a mobile node is roaming, in 6LoWPAN this would logically be the LoWPAN Edge Router. A PHA acts like a normal MIPv6 host, but additionally performs binding updates, HA tunneling and route optimization on behalf of other nodes. In order for a mobile node to use a PHA, it simply needs to perform a local binding update with (possibly much simpler) credentials, and to create a single tunnel to the PHA. The PHA architecture is show in Figure 4.5. This provides huge improvements in efficiency with regards to security associations and for route optimization which requires tunnels and other state for every correspondent node.

In order to use the PHA concept with 6LoWPAN, a mechanism for registering with the PHA would need to be defined for use inside the LoWPAN. The logical place to do this would be as an option for the 6LoWPAN-ND Node Registration message. Such an option would need to include the Home Agent's address or home prefix, the node's home address and some credentials (if L2 credentials are not sufficient). As the tunnel is local between the LoWPAN Edge Router and the LoWPAN Node, it could be realized as a simple IPv6 extension header option with very low overhead.

4.1.6 Proxy MIPv6

The network-based local mobility management (NETLMM) working group at the IETF [NETLMM] works on solutions for dealing with mobility locally within a domain, without requiring IPv6 nodes moving between points of attachment to change their IPv6 address or to implement MIPv6. In M2M systems this type of mobility between points of attachment within the same domain (controlled e.g. by an operator or enterprise) is quite common. As a solution to this problem space the WG has standardized *proxy MIPv6* (PMIPv6), which uses a local hierarchical structure of routers to handle mobility on behalf of nodes. PMIPv6 is specified in [RFC5213], with a problem statement documented in [RFC4830]. As discussed in Section 4.1.4, this model is more appropriate for use with 6LoWPAN than plain MIPv6.

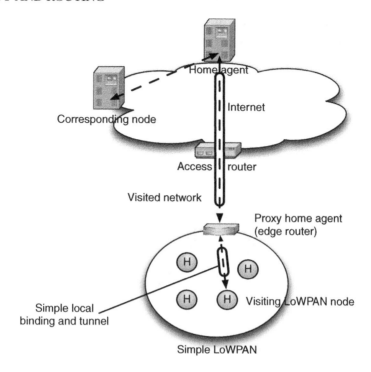

Figure 4.5 Example of a proxy Home Agent located on an edge router.

It allows LoWPAN Edge Routers or other local routers to proxy MIPv6 on behalf of attached LoWPAN Nodes.

The architecture of PMIPv6 is illustrated in Figure 4.6. The concept of a PMIPv6 domain is introduced, which is controlled by a *local mobility anchor* (LMA). The LMA function is usually combined with HA functionality. The LMA handles the local mobility of nodes with the help of *mobile access gateways* (MAGs), which are points of attachment supporting PMIPv6. MAGs send proxy binding updates to the LMA on behalf of mobile nodes attached to them. Using bidirectional tunnels built between each MAG and the LMA, the LMA is then able to forward traffic to the mobile node always using its static address (known as a mobile node home address). A binding in the LMA is made between this address and the temporary address from the visited MAG (the proxy care-of address). PMIPv6 makes use of RS/RA exchanges directly between the mobile node and MAG in order to detect when the mobile node has changed its point of attachment.

Although the PMIPv6 model would seem to fit well with 6LoWPAN, there are some problems that would still need to be solved:

- The RS/RA exchange defined in [RFC5213] is not compatible with a multihop Route-Over LoWPAN, and would require each LoWPAN Router to act as a MAG.

- PMIPv6 is meant to provide a separate 64-bit prefix for each mobile node.

- PMIPv6 only enables a node to talk with its point of attachment (default router), and requires NS/NA exchanges which are not required by LoWPAN Nodes otherwise using 6LoWPAN-ND.

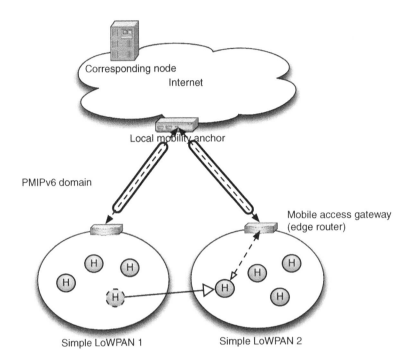

Figure 4.6 Example of PMIPv6 with 6LoWPAN.

4.1.7 NEMO

Network mobility (NEMO) is a solution for dealing with network mobility problems, when a router and the nodes attached to it, move their point of attachment all together. The philosophy behind NEMO is to extend Mobile IP so that each node does not need to run Mobile IP, instead only the router they are attached to runs Mobile IP. This philosophy fits the 6LoWPAN model perfectly as LoWPAN Nodes are not capable of dealing with MIPv6. Edge routers or other router entities run full IPv6 stacks, and have the capability of dealing with MIPv6.

The NEMO basic protocol is specified in [RFC3963]. The NEMO protocol works by introducing a new logical entity called the *mobile router*, which is responsible for handling MIPv6 functions for the entire mobile network. Mobile nodes which are part of the mobile network are called *mobile network nodes* (MNNs). These entities can be seen from Figure 4.7. MIPv6 normally only handles forwarding for the home addresses bound by mobile nodes. NEMO extends the functionality of the Home Agent to be able to deal with *prefixes* in addition to home addresses of mobile nodes. A mobile router functions like a normal MIPv6 host setting up a bidirectional tunnel with its Home Agent, but in addition it negotiates prefixes to be forwarded to it by the Home Agent. The Home Agent then forwards all packets matching the bound prefix (therefore packets for the MNNs) to the mobile router. A special flag in the binding update allows the mobile router to indicate it wants prefix forwarding, and a prefix option lets it configure prefixes with the HA. Alternatively, *prefix delegation* can be done using e.g. DHCPv6 [RFC3633, ID-nemo-pd].

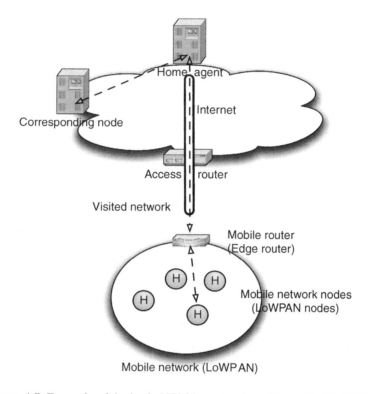

Figure 4.7 Example of the basic NEMO protocol working with 6LoWPAN.

NEMO is clearly beneficial when applied to mobile LoWPANs, where the entire LoWPAN including the edge router and associated nodes move together to a new point of attachment. When this happens the edge router acts as a NEMO mobile router. Using MIPv6, it binds its new care-of address in the visited network, and in addition the home LoWPAN prefix. Thus inside the LoWPAN no change can be noticed due to network mobility, as the LoWPAN continues to use the same prefix as in its home network. The HA takes care of forwarding all traffic destined for the LoWPAN prefix through the tunnel to the edge router and vice versa.

The drawback of NEMO is that it can not deal with individual node mobility on behalf of the LoWPAN Nodes. Thus a mobile LoWPAN Node would still have to implement MIPv6, unless it would use a proxy Home Agent or be moving within a PMIPv6 domain as discussed in the previous sections. Furthermore, NEMO starts to become complicated when mixing different kinds of node mobility along with PMIPv6.

4.2   Routing

IP mobility solutions consider techniques for preserving the IP address of a node as it moves from one point of attachment to another, along with methods for forwarding traffic to a node while roaming, and performing route optimization. IP routing on the other hand, deals with maintaining routing tables on IP routers which indicate which next-hop forwarding decision should be made for the destination of an IP packet. IP routing protocols maintain these routing tables using a wide variety of techniques ranging from ad hoc dynamic routing for wireless mesh networks to path vector inter-domain routing on the core Internet.

In this section we examine the basics of IP routing for 6LoWPAN, along with state-of-the-art techniques currently used or in development. Other techniques for forming mesh topologies below the IP layer, such as link-layer mesh and LoWPAN Mesh-Under techniques are not considered here, but were introduced in Chapter 2. When discussing IP routing for 6LoWPAN, there are two types of routing which need to be considered: routing inside a LoWPAN, and routing between a LoWPAN and another IP network. Routing is challenging for 6LoWPAN, with low-power and lossy radio links, battery-powered nodes, multihop mesh topologies, and frequent topology change due to mobility. Successful solutions must take the specific application requirements into account, along with Internet topology and 6LoWPAN mechanisms. Such a solution is being developed by the routing over low-power and lossy (ROLL) networks working group at the IETF. In this section we cover the requirements, route metrics, architecture and algorithm for ROLL. Relevant support mechanisms and routing algorithms have also been developed in the IETF mobile ad hoc network (MANET) working group.

First an overview of 6LoWPAN routing is given in Section 4.2.1, followed by consideration of how Neighbor Discovery fits into LoWPAN routing and border routing in Section 4.2.2. Routing requirements for embedded applications are presented in Section 4.2.3. Suitable routing metrics are discussed and summarized in Section 4.2.4. The ROLL architecture and algorithm basics are presented in Section 4.2.6. Algorithms developed by the MANET working group, including AODV, DYMO and OLSR, are discussed in Section 4.2.5. Finally the issue of border routing between LoWPANs and other networks is covered in Section 4.2.7.

4.2.1   Overview

As IP networks are *packet switched*, as opposed to circuit switched, forwarding decisions are made *hop-by-hop*, based on the destination address in a packet. Therefore reaching a destination node in a network from a source node requires building a path from the source to the destination node in route tables on nodes along the path. IP addresses are structured, and this structure is used to group addresses together under a single route entry. In IPv6 an address prefix is used for this purpose, which is why this is called prefix-based routing.

Routing and forwarding for 6LoWPAN were introduced in Section 2.5. In this present section we consider only IP routing performed at the network layer, and in particular IP routing algorithms useful for 6LoWPAN. Link-layer mesh techniques are transparent to IP, making the mesh look like a single link. IP routing in 6LoWPAN networks has some special characteristics due to the 6LoWPAN architecture:

- LoWPAN Routers typically perform forwarding on a single wireless interface, i.e. they receive a packet on their wireless interface from one node, and then forward it to the next-hop destination using the same interface. This is an important difference to how forwarding on IP routers normally works, where packets generally are forwarded between interfaces (and thus between links). The reason for this model is that typically not all nodes in a LoWPAN are reachable in a single wireless transmission, thus IP forwarding is used to provide full connectivity over multiple hops within the same "link".

- A LoWPAN has a *flat* address space, as all nodes in a LoWPAN share the same IPv6 prefix. This is due to the way 6LoWPAN compression is achieved, using the fact that all nodes in the network know common information to elide or compress fields. Therefore 6LoWPAN routing tables only contain entries to destination addresses in the LoWPAN, along with default routes.

- LoWPANs are stub networks, and are not meant to be transit networks between two different subnets. This simplifies the requirements for LoWPAN Routers.

Two kinds of routing can be performed with 6LoWPAN, *intra-LoWPAN* routing between LoWPAN Routers, and *border routing* performed at the edge of the LoWPAN by the LoWPAN Edge Router or an IPv6 router on the backbone link for Extended LoWPANs. These routing domains and the associated network-layer forwarding are illustrated in Figure 4.8 and Figure 4.9. Border routing will be discussed in Section 4.2.7.

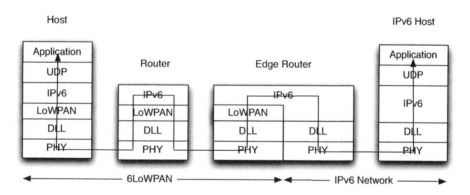

Figure 4.8 Stack view of forwarding inside the LoWPAN and across the edge router.

In addition to the above characteristics of the 6LoWPAN architecture, routing protocols also need to fulfill a large number of application, node and wireless-link-related requirements. These include the need to minimize energy usage, support node sleep cycles, achieve quality of service, support different addressing types (unicast, multicast, anycast), and work with mobility, all while using a minimal amount of memory and bandwidth. The challenge for routing protocols is that many of these requirements are conflicting, and thus for each application a balance needs to be found. Requirements developed in the ROLL working group for different commercial application domains will be presented in Section 4.2.3.

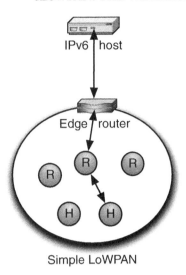

IPv6 host

Edge router

Simple LoWPAN

Figure 4.9 Topology view of forwarding inside the LoWPAN and across the edge router.

There are two main classes of routing protocols useful for 6LoWPAN: *distance-vector* routing and *link-state* routing. An analysis of existing distance-vector and link-state routing algorithms along with their applicability to wireless embedded applications is available in [ID-roll-survey].

Distance-vector routing: These algorithms are based on variations of the Bellman–Ford algorithm. Using this approach, each link (and possibly node) is assigned a cost, using appropriate route metrics. When sending a packet from node A to node B, the path with the lowest cost is chosen. The routing table of each router keeps soft-state route entries for the destinations it knows about, with the associated path cost. Routing information is updated either *proactively* (a priori) or *reactively* (on-demand) depending on the routing algorithm. Owing to their simplicity, low signaling overhead and local adaptive nature, distance-vector algorithms are commonly applied to 6LoWPAN.

Link-state routing: In this approach, each node acquires complete information about the entire network, called a graph. To do this, each node floods the network with information about its link information to nearby destinations. After receiving link-state reports from sufficient nodes, each node then calculates a tree with the shortest-path (least cost) from itself to each destination using e.g. Dijkstra's algorithm. This tree is used either to maintain the routing table in each node for hop-by-hop forwarding, or to include a source-route in the header of the IP packet. Link-state algorithms incur a large amount of overhead, especially in networks with frequent topology change. They require substantial memory resources for the amount of state needed by each node. Thus they are not suitable for distributed use among LoWPAN Nodes [ID-roll-survey]. Link-state algorithms may be usefully applied off-line on LoWPAN Edge Routers which have sufficient memory capacity if the signaling overhead for collecting the link-state information is reasonable. In this way only a single tree is constructed from the ER to nodes.

In order to update routing information throughout a network or along a path, routing protocols make use of either proactive or reactive signaling techniques. These terms can be defined as:

Proactive routing: Algorithms using a proactive approach build up routing information on each node before the routes are needed. Thus they proactively prepare for the data traffic by learning routes to all possible or likely destinations. Most protocols that are used in inter-domain or intra-domain IP routing use a proactive approach as topologies are stable. Examples of proactive algorithms can also be found from MANET, for example optimized link-state routing (OLSR) [RFC3626] and topology dissemination based on reverse-path forwarding (TBRPF) [RFC3684]. The advantage of this approach is that routes are immediately available when needed, but this comes at the cost of increased signaling overhead especially with frequent topology changes and increased state for routers.

Reactive routing: Reactive routing protocols store little or no routing information after autoconfiguration of the routing protocol. Instead, routes are discovered dynamically only at the time they are needed. Thus a process called *route discovery* is executed when a router receives a packet to an unknown destination. Examples of reactive algorithms include the MANET ad hoc on-demand distance vector (AODV) [RFC3561] and dynamic MANET on-demand (DYMO) [ID-manet-dymo] protocols, along with the ZigBee routing algorithm derived from AODV [ZigBee]. The advantage of this approach is that signaling and route state grows only as needed, and it is especially well suited to ad hoc networks with frequent topology change and mainly peer-to-peer communications.

Advanced techniques which may be applied in 6LoWPAN routing protocols include constrained routing using compound route metrics, local route recovery, flow labeling to achieve Multi Topology Routing (MTR), forwarding on multiple paths with multipath routing, and traffic engineering. Several of these techniques are being considered for the ROLL routing algorithm.

4.2.2 The role of Neighbor Discovery

As IPv6 makes use of the Neighbor Discovery protocol for interactions between neighbors, it is an integral part of IPv6 networking. How does Neighbor Discovery interact with IPv6 routing protocols? First and foremost both IPv6 [RFC4861] and 6LoWPAN [ID-6lowpan-nd] Neighbor Discovery take care of next-hop decisions for hosts, along with the maintenance of neighbor cache, destination cache and router cache information. ND is instrumental in bootstrapping onto a network and maintaining information between neighbors. Furthermore ND plays an important role in Neighbor Unreachability Detection and the ability to let senders know about alternative next-hop routers. All IPv6 routing protocols make use of ND information and often signaling in addition to messages specific to the routing protocol. We can think of ND as a "single-hop protocol", whereas routing protocols take care of multihop information. Neighbor Discovery is covered in detail in Section 3.2.

ND can be used as a *proxy service* between two links [RFC4861, RFC4389]. In this case ND acts as a dynamic single-hop routing protocol, sharing information about nodes on one interface with nodes on another interface. In 6LoWPAN-ND [ID-6lowpan-nd], edge routers perform a similar function in Extended LoWPAN topologies, between their 6LoWPAN wireless interface and their backbone link interface. Here forwarding decisions between the LoWPAN and backbone link interfaces of the edge router can be made using ND information stored in the whiteboard of the edge router. As the whiteboard contains bindings for all the IPv6 addresses registered by nodes in the LoWPAN, it only forwards traffic to and from nodes in the network. Although ND can help with forwarding across edge routers in Extended LoWPANs, it does not help with intra-LoWPAN routing or with border routing between the Extended LoWPAN and routing protocols running in the local domain.

4.2.3   Routing requirements

When considering a routing protocol for use with 6LoWPAN, it is important to take the requirements for a wide range of embedded system applications into account. Typical Internet applications use a client–server approach between PCs and web servers, making requirements fairly easy to analyze. As discussed earlier, wireless embedded applications have a wide range of often conflicting requirements. Compounded with the 6LoWPAN architecture, wireless link-layer technology and node limitations, these requirements are difficult to fulfill. In the ROLL working group, a large effort has been made on the analysis of routing requirements for four major application areas:

Urban networks: In the future, wireless embedded networks will be deployed widely in urban environments for purposes such as environmental monitoring, automatic meter reading and smart grid. Routing requirements for urban applications have been collected in [RFC5548]. In order for a routing protocol to be useful for this application space, it needs to be energy-efficient, scalable and autonomous while at the same time taking into account the limited capabilities of nodes.

Industrial networks: The deployment of low-cost wireless field devices in industrial facilities will increase safety and productivity. The requirements for a routing protocol used in the industrial application space have been presented in [ID-roll-indus]. These requirements are influenced by the ISA100 standard [ISA100.11a]. Although ISA100 currently uses a proprietary centralized routing method, it is understood to need a standard IETF routing solution in the future. The three major requirements for wireless embedded networking in this application space are low power, high reliability and ease of installation and maintenance.

Building networks: Commercial building automation is an important application area for wireless embedded networking, such as the facility management example introduced in Section 1.1.5. The routing protocol requirements for this application space have been analyzed in [ID-roll-building]. There is a long tradition of installing automation technology in buildings using wired technology, more recently using modern application protocol standards such as the building automation control network (BACnet) and open building information exchange (oBIX), which we cover in Section 5.4.7. Low-power wireless technology is making a breakthrough in this area, and there is a great interest in end-to-end IP solutions. Important requirements for a successful protocol in this space include autoconfiguration, manageability, scalability and low-energy use for battery-powered devices.

Home networks: A large number of applications of low-power wireless networking in the home have been identified, including home healthcare, automation, security, energy monitoring and entertainment. Requirements for a routing protocol in this environment are analyzed in [ID-roll-home]. Unlike other application areas analyzed in ROLL, this space is consumer oriented, placing a different emphasis on requirements. Devices are cost sensitive, while at the same time required to be physically small with a long battery life. Important requirements include peer-to-peer communications, zero-configuration and adaptability to link changes.

A non-exhaustive list of mandatory requirements identified in these ROLL documents have been summarized in Table 4.1. The requirements are categorized by their general type, and grouped together to easily identify where applications have common requirements. In addition to these MUST requirements, a large number of SHOULD requirements have also been identified in the documents. Note that we use IETF-style key words (MUST, SHOULD) in this table [RFC2119]. Although these have been identified to aid in the design of the ROLL routing protocol, they are also very useful for understanding these application spaces, and for considering the use of other routing protocols with 6LoWPAN.

It should be kept in mind that ROLL aims at routing for more general IP networks as well as 6LoWPAN. Therefore not all requirements necessarily apply to 6LoWPAN, although the vast majority do. The 6LoWPAN working group has made an analysis of routing requirements related to 6LoWPAN specifically in [ID-6lowpan-rr]. In this document the device, link and network properties are analyzed along with security considerations.

4.2.4 Route metrics

In the path selection process of a routing protocol, *route metrics* are used to choose the best route. Typical metrics for IP routing include e.g. hop count, bandwidth, delay, MTU, reliability and sometimes load. As seen in the previous section on routing requirements, 6LoWPAN networks have specific requirements compared to traditional wired or even mobile ad hoc networks. Routing metrics employed in 6LoWPAN routing may be very specific to the application where they are being used. Route metrics for use by the ROLL working group have been specified in [ID-roll-metrics].

In most 6LoWPAN networks, such as for the applications considered in ROLL, different types of metrics are considered which can be classified as follows:

- link versus node metrics

- qualitative versus quantitative metrics

- dynamic versus static metrics

Examples of link metrics include throughput, latency and link reliability. Node metrics may include e.g. memory, processing load and residual energy. The metrics identified in [ID-roll-metrics] are summarized in Table 4.2.

Table 4.1 Non-exhaustive summary of requirements identified for ROLL.

Type	Areas	Requirement
Addressing	U, I, B, H	The protocol MUST support unicast, anycast and multicast addressing.
Addressing	B, H	A device MUST be able to communicate peer-to-peer with any other device in the network.
General	B	The protocol MUST support the capability of nodes to act as a proxy for sleeping nodes. The proxy stores packets for a sleeping node, and delivers them during the next awake cycle.
Traffic flow	U, I	The protocol MUST support multiple paths to a given destination for reliability and load balancing.
Configuration	U, I, B, H	Autoconfiguration of the routing algorithm MUST be supported, without human intervention.
Configuration	B	It MUST be possible to commission devices without requiring any additional commissioning devices (e.g. a laptop).
Configuration	U, I, B, H	The protocol MUST be able to dynamically adapt to changes based on network layer and link-layer abstractions.
Configuration	I	The protocol MUST support the distribution of configuration information from a centralized management controller.
Management	H	The protocol MUST support the ability to isolate a misbehaving node.
Scalability	U	The protocol MUST be able to support a large number of nodes in regions containing on the order of 10^2 to 10^4 nodes.
Scalability	U	The protocol MUST accommodate a very large and increasing number of nodes without deteriorating selected performance parameters.
Scalability	B	The protocol MUST be able to support networks with at least 2000 nodes, and subnetworks with up to 255 nodes each.
Scalability	H	The protocol MUST support up to 250 devices in the network.
Performance	I	Success or failure regarding route discovery MUST be reported within several minutes and SHOULD be reported within tens of seconds.
Performance	H	The protocol MUST converge within 0.5 seconds with no mobility, respond to topology change within 0.5 seconds if the sender has moved, and with 2 seconds if the destination has moved.
Performance	H	Sleeping nodes MUST be taken into account by the routing algorithm.
Metrics	U, I, B, H	The protocol MUST support different link and node metrics for use in constraint-based routing.
Security	U, I, B	The protocol MUST support authentication and integrity measures, and SHOULD support confidentiality measures.
Security	U	Trust MUST be associated between a node and the network before autoconfiguration or routing participation can occur.
Security	H	The protocol MUST allow for low-power, low-cost network devices with limited security needs.
Security	H	Protection against unintentional inclusion in neighboring networks MUST be provided.

U = Urban, I = Industrial, B = Building, H = Home.

Table 4.2 Routing metrics identified for ROLL.

Metric	Type	Description
Node memory	QT, ST	The memory available for routing information on a node.
Node CPU	QT, ST	Computational power, not considered to be critical in most ROLL applications.
Node energy	QT, DY	The residual energy left for battery-powered nodes, important for optimizing network lifetime.
Node overload	QT, DY	A simple indication of the network load (e.g. queue size) of a node.
Link throughput	QT, DY	The total and currently available throughput of a link.
Link latency	QT, DY	The range of latency and current latency for a link.
Link reliability	QT, DY	The link reliability specified as e.g. average packet error rate, which is a critical routing metric.
Link coloring	QL, ST	This static attribute is used to prefer or avoid specific links for specific traffic types.

QT = Quantitative, QL = Qualitative, ST = Static, DY = Dynamic.

Most of these metrics are used for building and maintaining the routing topologies, others for making forwarding decisions (as we will discuss in Section 4.2.6) whereas some are also applied to constraint-based routing. The ROLL routing protocol, for example, will make use of a very granular *depth* (weighted hop count) metric for building the basic topology. Additionally, application-specific sets of metrics are then used for routing on the topology.

The use of dynamic metrics is particularly challenging, as it may lead to routing instability. Furthermore the reporting frequency should be minimized, and multi-threshold techniques applied to dynamic metrics to minimize signaling overhead. Finally, as a set of metrics will be used for path calculation, it is important that metric calculation is consistent throughout the same routing domain.

4.2.5 MANET routing protocols

The mobile ad hoc network (MANET) working group at the IETF was formed in 1997, with the goal of understanding the requirements and designing solutions for wireless ad hoc IP-based network applications [MANET]. The WG has produced a large variety of routing protocols which can be categorized by their scheme for dealing with route updates, proactively or reactively, and by their routing technique, distance-vector or link-state. These terms were defined earlier in Section 4.2.1. Protocols developed in MANET are mainly meant for routing in ad hoc mobile networks using WLAN technology, where the majority of traffic is peer-to-peer. Applications include nomadic computing, emergency and rescue networks, and military applications. In this section three common MANET protocols are introduced: the ad hoc on-demand distance vector (AODV), dynamic MANET on-demand (DYMO) and optimized link-state routing (OLSR) protocols.

In addition to routing protocols, MANET has also produced valuable work on basic mechanisms for supporting routing in these environments. A common packet format for use by all MANET protocols has recently been developed in [RFC5444]. As an addition to this, a standard time representation has been defined in [RFC5497]. Finally, a *two-hop* neighborhood discovery protocol is currently being developed for use in collecting route information in a standard way [ID-manet-nhdp].

The protocols and mechanisms developed in MANET can be applied to 6LoWPAN networks. They are especially useful for applications with very similar requirements to typical MANET applications. The biggest challenge for applying a MANET algorithm to 6LoWPAN is in reducing the overhead of signaling packets and simplifying the algorithms. The common MANET protocol format and two-hop ND can also be usefully applied to new routing protocols for 6LoWPAN. We should keep in mind that MANET protocols are not really applicable to many wireless embedded system applications, for example those examined by ROLL and introduced in Section 4.2.3. These systems require Internet-connected networks with strong ER-to-node and node-to-ER traffic flows, along with the need for multipath routing and often traffic engineering [ID-roll-survey].

AODV

The ad hoc on-demand distance vector (AODV) protocol enables mobile ad hoc multihop networks by quickly establishing and maintaining routes between nodes, even with quickly changing (dynamic) topologies. AODV creates routes to destinations when needed for data communications (reactive), and only maintains actively used routes. It includes methods for local repair, and includes a destination sequence number to ensure loop-free operation. AODV is purely a route table management protocol; after routes have been established they are simply used by IP for forwarding.

A small set of messages are used for discovering and maintaining routes by AODV and similar protocols. A *route request* (RREQ) is broadcast throughout the network in order to find paths to a destination. This is responded to with a *route reply* (RREP) by an intermediate router or by the destination. Figure 4.10 shows an example of reactive route discovery and forwarding in an ad hoc network. The *route error* (RERR) message is used to notify about broken links along a path. These messages are sent over UDP, one hop at a time, between the AODV processes running on ad hoc routers. AODV is specified in detail in [RFC3561].

As AODV was the first reactive distance-vector routing algorithm standardized in the IETF, it has been used as a model for many other routing algorithms. For example, the routing algorithm that ZigBee used in its network layer designs is modeled on the AODV algorithm with modifications to minimize overhead and to function on MAC addresses rather than IP addresses.

DYMO

A new reactive distance-vector routing protocol called the dynamic MANET on-demand (DYMO) protocol has been developed in the MANET WG [ID-manet-dymo], making improvements on previous protocols such as AODV and dynamic source routing (DSR) [RFC4728]. This protocol makes use of the same types of route discovery and maintenance messages as AODV. The main improvements compared to previous work include:

- improved convergence in dynamic topologies
- use of the common MANET packet format [RFC5444]
- support for a wide range of traffic flows

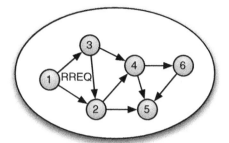

1. RREQ for node 5 broadcast over multiple hops.

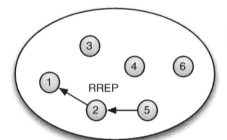

2. RREP unicast back to node 1, creates route entries.

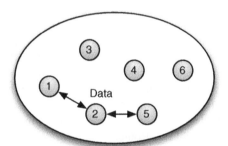

3. Route entries in 1, 2 and 5 enable forwarding.

Figure 4.10 Example of reactive distance-vector routing.

- consideration for Internet interconnectivity

- takes hosts and routers into account

Proposals have been made in the 6LoWPAN working group for the adaptation of DYMO for use as a LoWPAN RFC4944 mesh routing algorithm. The standardization of such link-layer algorithms is however not currently on the agenda in the IETF. Instead, MANET protocols for IP routing can be applied with optimizations to 6LoWPAN directly.

OLSR

The MANET WG has also produced a proactive link-state routing protocol called the optimized link-state routing (OLSR) algorithm. Originally specified in [RFC3626], an improved OLSRv2 has been developed in the working group in [ID-manet-olsrv2]. This algorithm applies optimization to the classical link-state algorithm for use in mobile ad hoc networks. In order to build link-state tables, OLSR routers regularly exchange topology information with other routers. The flooding of this information is controlled by the use of selected multipoint relay (MPR) nodes. These MPR nodes are used as intermediate routers, and thus enable a kind of clustering technique. The OLSR algorithm makes use of the standard MANET packet format and two-hop ND techniques.

OLSR is best suited for relatively static ad hoc networks, thus minimizing the number of link-state updates throughout the network, which can cause a lot of overhead. OLSR is not very well suited to 6LoWPAN Routers because of the large amount of signaling and routing state. Link-state protocols are also not well suited to tree topologies, as often found in 6LoWPAN applications. Link-state approaches like OLSR may have possible uses for the partial optimization of larger route topologies or off-line use on border routers, for example with the ROLL routing algorithm.

4.2.6   The ROLL routing protocol

The routing over low power and lossy (ROLL) networks working group was formed in the IETF to analyze the requirements for and standardize a routing protocol for embedded applications such as urban ubiquitous networks, industrial automation, building automation and home automation [ROLL]. *Low-power and lossy networks* (LLNs) are typically made up of embedded devices with limited processing, memory and power resources. A wide range of link-layer technologies are aimed at by the ROLL WG, including IEEE 802.15.4, Bluetooth, low-power WiFi, along with low-power and lossy wired and power line communication (PLC) technologies. The working group focuses only on routing for general IPv6 and 6LoWPAN networks, with IPv4 support out of scope. For this reason the terminology used in ROLL differs from 6LoWPAN terminology, for example LLN instead of LoWPAN and LLN border router instead of LoWPAN Edge Router.

The applications and link-layers considered by ROLL have several specialized properties:

- Traffic patterns are not only peer-to-peer unicast flows, but more often point-to-multipoint or multipoint-to-point flows. Most applications of LLNs are Internet connected.

- Routers in LLNs have a very small, hard bound on state (limited memory).

- Most LLNs must be optimized for energy consumption.

- In most cases LLNs will be deployed over links with a limited frame size.

- Security and manageability are extremely important as LLNs are typically autonomous.

- The application spaces aimed at by ROLL are heterogeneous. Each may need a different set of features along with routing metrics to fulfill its requirements.

The ROLL WG started out with a requirements analysis conducted on the main application areas ROLL is aiming at, the results of which were covered in Section 4.2.3. Based on these requirements a survey was made of existing IETF routing protocols. It was concluded that it is impossible for any of these to be used for ROLL purposes without major modifications [ID-roll-survey]. The working group is now focusing on generating actual specifications for a routing protocol and at the time of writing ROLL has produced initial documents for its work items. Note that many of these documents are still an early work in progress – see the IETF ROLL web pages for the latest information [ROLL]. The following activities are going on in the ROLL working group:

Metrics: The routing metrics useful for path calculation have initially been specified in [ID-roll-metrics]. These metrics were presented in Section 4.2.4. In practice, work will still need to be done on algorithms for *evaluating* appropriate metrics for each specific application space.

Architecture: The basic architectural requirements for ROLL have been captured in the requirement documents. Terminology for use in ROLL has been specified in [ID-roll-terminology].

Security: A security framework is being developed in the working group. An overview of requirements and some techniques for ROLL security have been provided in [ID-roll-security], and some considerations for trust management are collected in [ID-roll-trust].

Protocol: The goal is to design a routing protocol that can be successfully applied to fulfill the routing requirements of the four application areas identified for ROLL. Early contributions towards this goal have been made, and at the time of writing the initial ROLL protocol is being designed.

The rest of this section gives an overview of the ROLL terminology, architecture and basic routing concepts based on the current contributions to the standardization effort. This overview of the basic ROLL protocol and advanced options is based on the LLN routing fundamentals proposal [ID-roll-fundamentals], and is meant to give a general overview of the concept.

ROLL architecture

The architecture of LLNs is very different from that considered by MANET protocols or in research work done on wireless sensor networking. In fact it has more in common with traditional intra-domain IP routing methods. The ROLL protocol can be classified as a proactive distance-vector algorithm with advanced options for constraint-based routing, multi-topology routing and traffic engineering. The key requirements or assumptions for LLNs that affect the routing architecture are:

- LLNs are Internet-connected stub networks, with support for multiple points of attachment (multiple border routers to other IP networks).

- The majority of traffic flows are going to or from border routers using unicast, point-to-multipoint or multipoint-to-point flows. Node-to-node communication is less common, but may require specific constraints.

- Support for dynamic topologies and mobility is required.

- Support is required for multipath routing, and thus multiple forwarding options.

- Support is required for multiple node and route metrics, and their application in constraint-based and multi-topology routing. The evolution of metrics and support of multiple scenarios are important.

- A course-grained depth metric is assumed for general use, which is independent of the specific scenario. It is not assumed that this metric provides absolute loop avoidance.

- Routers in LLNs have limited memory resources.

- Most applications will require enterprise-class security.

The ROLL architecture is presented in Figure 4.11. The generalized routing architecture is similar to that of an Extended LoWPAN defined by 6LoWPAN. Routers with interfaces to the LLN and another IP link are called *LLN border routers* (LBRs). There may be several LBRs connecting an LLN to a backhaul or backbone link. The LBR functionality is typically implemented on the same device as LoWPAN Edge Router functionality. Inside the LLN, the network is made up of LLN routers and LLN hosts. Hosts do not participate in the LLN routing algorithm, and instead simply choose default LLN routers. The addressing in an LLN can either be based on IPv6 prefixes for use with general IPv6 or on destination addresses for use with 6LoWPAN. The ROLL routing protocol operates within the LLN domain, and terminates at the LBR. Border routing solutions across the LBR are discussed in Section 4.2.7.

The base of the routing protocol uses a *graph* structure between nodes and LBRs, which can be seen from Figure 4.11. This basic topology needs to be discovered and maintained using a minimal amount of signaling. After the basic topology is constructed, the routing protocol maintains upstream (from node to LBR) and downstream (from LBR to node) paths. Forwarding along these paths is then performed using IPv6 forwarding. The coordination of constrained routing, multi-topology routing and traffic engineering typically needs to be performed from a centralized place in the network, for which LBRs are a logical place. Other options may be performed in a distributed manner, such as node-to-node optimized routes.

There are two concepts important to understanding ROLL protocol operation:

Metric granularity: ROLL uses the concept of a very granular (16–32 values) route metric called *depth*. This metric is used by the basic ROLL protocol mechanisms for building the graph, making use of siblings and for loop avoidance. The evaluation of depth is simple for all routers and nodes and is independent of the application scenario. In addition, fine-grained sets of metrics (such as those presented in Section 4.2.4) along with evaluation algorithms are applied in an application-specific manner to achieve routing on the basic graph structure.

Routing time scale: ROLL makes routing decisions on two different time scales: route-setup time and packet-forwarding time. In *route-setup time* the routing protocol maintains the basic graph topology and routing tables using static or slowly moving metrics, which is a continuous process. In addition, ROLL enables *packet-forwarding time* decisions to be made using dynamic metrics on a packet-by-packet basis, for example the immediate use of alternative next-hop routers upon failure.

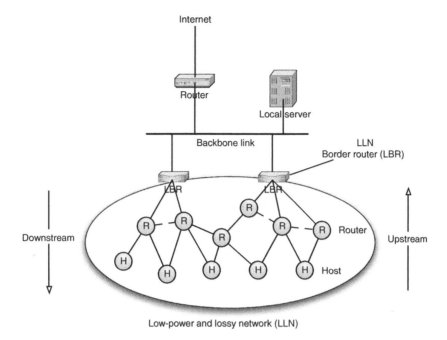

Figure 4.11 The ROLL architecture.

Building and maintaining topology

After nodes perform commissioning, the first process in an LLN is to build up an underlying routing topology. In a topology with flows between nodes and LBRs, this can be achieved by forming a graph pointed towards the exit points of the network (LBRs). Nodes in IPv6 networks already send control signaling in the form of Neighbor Discovery messages. These messages can naturally be used to build up a graph using specific rules. In practice, graph information includes e.g. depth, path cost, sequence number and lifetime, and is periodically sent from LBRs down the network towards the furthest nodes. Using specific rules, routers are able to attach to the graph by choosing a set of default next-hop routers (parents or siblings) towards LBRs. This topology automatically builds up multiple distance-vector paths from nodes up to LBRs, which can then be used for node-to-LBR packet forwarding. Such a topology allows for nodes to change their position in the topology, for topologies to be merged, and for topologies which are not grounded on an infrastructure (ad hoc) to operate. The ROLL topology is maintained using periodic advertisements (e.g. ND Router Advertisements could be used) sent from LBRs towards nodes, then propagated by each router. By using existing ND signaling, the protocol can avoid the overhead associated with routing topology maintenance.

In order for the routing protocol to maintain downstream routing information, nodes are required to *disseminate* path cost information upstream towards LBRs. You can consider this as painting the graph topology already built with additional downstream routing information.

Node dissemination can be achieved using ND messages, data piggybacking or specialized signaling. Disseminated routing information can be used to maintain distance-vector information in intermediate routers or to record reverse routes. In this way distance-vector state in routers can be bounded using both techniques. Finally, node route dissemination is important for LBRs to build up global information about the LLN for use in multi-topology routing and traffic engineering.

Forwarding traffic

The basic topology maintained by ROLL enables forwarding between LLN nodes and LBRs (upstream). The second mechanism of node route dissemination up the graph towards LBRs then enables forwarding from LBRs towards LLN nodes (downstream) along with basic node-to-node routing along the graph. Reliability is improved by providing multiple routing paths to nodes by using multiple default routers chosen at packet-forwarding time. Upstream and downstream forwarding are illustrated in Figure 4.12.

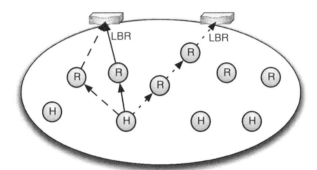

Upstream routing, three default routers

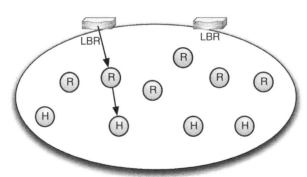

Downstream routing

Figure 4.12 Examples of upstream and downstream forwarding with ROLL.

When a node wants to send traffic to an off-link destination, it does so by sending it first to its best next-hop default router. If forwarding through this default route fails, secondary default routers can be tried without indicating failure to the application process (note, this is an exception to normal IP next-hop determination). Default routers are ranked as part of topology discovery and maintenance at route-setup time. The decision for which next-hop router to use from the list of available ones is made at packet-forwarding time using suitable metrics and possible forwarding constraints.

Downstream forwarding can be achieved using hop-by-hop distance-vector state, source routing or a combination of both (loose source routing). This depends on the amount of memory available in routers and the optional features employed by the LBRs. Distance-vector downstream forwarding is a basic default as this state is automatically available due to node dissemination. Source routing can be achieved either using stateless source routing, where the route is specified by the LBR with an IPv6 routing header, or using states at each node and flow labeling to indicate the path.

Optimizations

The ROLL routing protocol is based on the basic ROLL topology presented above, along with node route dissemination. These simple but robust techniques enable both multipath upstream forwarding and downstream forwarding using multiple techniques. To achieve many of the requirements identified in ROLL, additional features may be needed. Solutions being considered in ROLL include:

Traffic engineering: For applications with requirements for particular quality of service, off-line traffic engineering may be required to fulfill the requirements, for example in industrial automation. This requires LBRs with sufficient resources to collect sufficient information about the LLN to engineer optimizations. This can be achieved by assigning labels and using multi-topology routing, or by installing source routes to specific nodes.

Node-to-node flows: Node-to-node forwarding is achieved with basic ROLL features using intersecting routers up the graph with distance-vector state for the destination, which in the worst case is the LBR. For applications with node-to-node flows, optimization can be achieved using *boulevard routes* between routers at the same depth, or by utilizing a reactive distance-vector feature.

Mobility support: Support for node mobility with MIPv6 [RFC3775], network mobility with NEMO [RFC3963] or the combined mobile ad hoc network mobility (MANEMO) may be useful for many applications in addition to the ROLL algorithm. This integration of mobility solutions with ROLL is an important optimization to be considered. Mobility solutions for 6LoWPAN were discussed in Section 4.1.

4.2.7 Border routing

Simply routing within a LoWPAN is not very useful for the majority of Wireless Embedded Internet applications, where most traffic flows are coming from the Internet towards one or more LoWPAN Nodes, or from LoWPAN Nodes towards the Internet. As a result, the ROLL protocol is being specified to optimize for these types of flows through LLN border routers, which in 6LoWPAN are usually also edge routers. How does the border router maintain route entries between its interfaces, which may belong to two different routing domains? If there are different routing protocols on these interfaces, how is information shared between them? This section answers these questions, introducing border routing and solutions for 6LoWPAN.

Border routing between two IP routing domains is a common issue on the Internet, where intra-domain and inter-domain routing protocols intersect. Traditionally it is less common on the very edge of the Internet, as local access technologies such as WiFi and Ethernet use bridging techniques to connect devices to the Internet. With the advent of IP routing in wireless stub networks using mesh routing protocols such as those presented in the previous section, border routing becomes an issue here as well. In 6LoWPAN we have three border routing cases to consider:

- Simple LoWPAN

- Extended LoWPAN

- route redistribution

In Simple LoWPANs there is only a single edge router for the LoWPAN, and the subnet prefix of its LoWPAN interface is different from that of its IPv6 interface. As the Simple LoWPAN and the IPv6 link are on different subnets, prefix-based route entries between the two prefixes are used. In addition, [ID-6lowpan-nd] mandates that edge routers must filter out any outgoing traffic from or incoming traffic to addresses not in its whiteboard. Therefore it is not necessary to run a routing protocol on the edge router's IPv6 interface. A routing protocol may be needed over some access networks, or particular deployment strategies may result in a large number of Simple LoWPANs with backhaul links on the same prefix (e.g. over GPRS). Figure 4.13 shows an example of border routing in the Simple LoWPAN case, with ROLL routing in the LoWPAN and OSPF routing on the backhaul link. The route to 2001:4fa2:0001::/48 would be *redistributed* to OSPF.

Border routing between an Extended LoWPAN and an IP network can be performed in two different places. As the LoWPAN interfaces and IPv6 interfaces of edge routers in Extended LoWPANs are in the same subnet, routing between them must be done using destination (exact match) route entries. The simplest way to achieve this is to use edge router whiteboard entries to maintain these route table entries. A routing protocol could be used on the backbone link to enable routing over the backbone between parts of the Extended LoWPAN. Border routing can also be performed on the router between the backbone link and another IP network. Routing here is achieved using prefix-based route entries, just as in the Simple LoWPAN case.

If border routing involves routing algorithms on both of its interfaces, it may be necessary to perform *route redistribution* between those interfaces. In route redistribution, a router advertises some routes maintained by an algorithm on one interface into the algorithm of another interface. As LoWPANs are stub networks, this redistribution always happens from the LoWPAN routing protocol to the protocol on an IP interface. Candidate routing protocols useful with border routers in combination with the ROLL algorithm include OLSR and the open shortest path first (OSPF) [RFC2328] protocol.

4.3   IPv4 Interconnectivity

Although IPv6 support is becoming more popular on the Internet, IPv4 is still being utilized for the vast majority of Internet data, and for many years there will still be swatches of

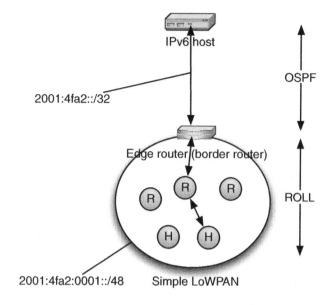

IPv6 host

OSPF

2001:4fa2::/32

Edge router (border router)

R

R

R

ROLL

H

H

2001:4fa2:0001::/48   Simple LoWPAN

Figure 4.13 Border routing example.

the Internet where only IPv4 packets can be forwarded. 6LoWPAN by nature is IPv6, with all LoWPAN Nodes and edge routers acting as IPv6 hosts or routers inside a purely IPv6 (6LoWPAN) network. Therefore the interconnection of 6LoWPAN over IPv4 networks is an important consideration for the deployment of 6LoWPAN on a wide scale. This section examines when IPv4 interconnectivity is required, introduces how 6LoWPAN networks can be integrated into the global IPv4 network and gives an overview of common IPv6-in-IPv4 tunneling techniques.

4.3.1 IPv6 transition

Because 6LoWPAN networks purely utilize IPv6, and embedded applications using 6LoW-PAN are specialized, integrating them with the IPv4 Internet is fairly straightforward. The majority of problems with IPv4-IPv6 transition on the Internet involve PCs, Internet service providers (ISPs) and servers and have to do with:

1. IPv4 nodes communicating with IPv6 nodes

2. IPv6 nodes communicating with IPv4 nodes

3. IPv6 nodes communicating over an IPv4 network

4. IPv4 nodes communicating over an IPv6 network

5. dual-stack nodes communicating over IPv4 or IPv6 networks

The large combination of different IPv4/IPv6 interactions has made the transition to IPv6 a slow process, especially for ISPs. In the case of 6LoWPAN we do not need to consider dual-stack hosts or IPv4 hosts, which simplifies the problem down to IPv6-only (LoWPAN) nodes communicating with IPv4-only nodes or with other IPv6-only nodes across IPv4 networks.

Typical applications involve 6LoWPAN Nodes communicating with dedicated server applications, either located on the same network or remotely on the Internet. If the server is located on the same backbone link (e.g. Ethernet) or IPv6 network, then IPv6 is used directly between LoWPAN Nodes and IPv6 nodes. It is common for industrial automation, building automation and asset management applications to use this type of system topology.

Applications in M2M communications, remote monitoring or smart metering may require LoWPAN Nodes to communicate with remote servers. In the majority of cases, these servers can easily be enabled with IPv6 support which simplifies the problem space for 6LoWPAN IPv4 interconnectivity down to LoWPAN Nodes and IPv6 nodes communicating over IPv4 networks. Such systems typically provide a web page or web service interface on the server for end-users and other systems, which is then independent of IP version.

The IETF has developed a large range of IPv6 transition mechanisms, which have lately been narrowed down to a few techniques that have been shown to work in practice over the past decade. The following techniques are most commonly applied today for enabling IPv6 communications over an IPv4 infrastructure:

Dual-stack: This is a technique for providing complete support for both Internet protocols in hosts and routers, defined in [RFC4213]. As 6LoWPAN Nodes are IPv6-only, this technique is not applicable. Edge routers may very well be configured as dual-stack hosts, but it does not help IPv4 interconnectivity for the whole LoWPAN.

Configured tunneling: IPv6-in-IPv4 tunneling (also known as 6in4) is employed by creating point-to-point tunnels across IPv4 networks, specified in [RFC4213]. Configured tunneling is used by a host or router to acquire a global IPv6 prefix. This technique is especially common in administered networks, and is useful for providing IPv6 for a Simple LoWPAN or Extended LoWPAN, or for remote hosts.

Automatic tunneling: Automatic tunneling (also known as 6to4) uses IPv6-in-IPv4 tunneling, automatically finding the end-point using the IPv4 anycast address 192.88.99.1. Using this technique specified in [RFC3056] a well-known 6to4 router must be used. The technique is useful for the same purposes as configured tunneling.

Next we look at both automatic and configured IPv6-in-IPv4 tunneling techniques in more detail.

4.3.2   IPv6-in-IPv4 tunneling

Tunneling is a common technique for providing IPv6 connectivity across an IPv4 infrastructure. Such an IPv6-in-IPv4 tunnel is achieved by encapsulating IPv6 packets within IPv4, thus using IPv4 as the link layer for IPv6. These encapsulated IPv6 packets are indicated by IP protocol number 41. There are also other schemes for IPv6 tunnel encapsulation, for example UDP encapsulation which is used for traversing NATs which block protocol 41 traffic. When packets reach the other end-point of the tunnel, the IPv6 packet is decapsulated and processed. MTU determination for a tunnel is important, as the IPv4 header adds a 20-byte overhead, thus the MTU for a tunnel is typically set to between 1280 and 1480 bytes. Dynamic MTU determination can also be employed using IPv4 *path MTU discovery* [RFC1191].

The end-points of tunnels can be either hosts or routers. In *automatic tunneling*, one end-point is always a well-known router, and the other can be either a router or a host. In *configured tunneling*, the end-points both tend to be routers as the tunnel requires configuration and maintenance, although also hosts can act as end-points. When applying tunneling to 6LoWPAN, we are interested in router-to-router tunneling to either the LoWPAN Edge Router or an IPv6 router in the local domain.

Automatic tunneling [RFC3056] is already widely deployed on the Internet, with support included in most operating systems. Well-known 6to4 tunnel end-points have been configured around the world for interconnecting IPv6 nodes to the IPv6 Internet over IPv4 infrastructure. The nearest 6to4 tunnel end-point is found using the IPv4 anycast address 192.88.99.1. You can check your nearest one by pinging or tracing a route to that address. Some problems with automatic tunneling include NATs blocking IP protocol 41, the difficulty of

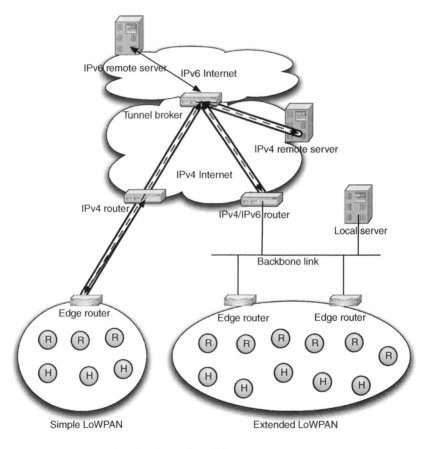

Figure 4.14 Configured IPv6-in-IPv4 tunneling example.

debugging automatic tunnel end-points and the inability to control and authenticate the end-points. Recent improvements to automatic tunneling include the intra-site automatic tunnel addressing protocol (ISATAP) [RFC5214] and 6to4 "rapid deployment" [ID-despres-6rd].

Configured tunneling [RFC4213] offers some advantages over 6to4 as the tunnel end-points can be fully configured, tunnel brokers [RFC3053] can be employed and alternative encapsulation techniques are easier to support. Figure 4.14 shows an example of 6LoWPAN networks and configured tunneling. The most common way of using configured tunneling is by employing a tunnel broker, which can be a public service such as http://www.freenet6.net or a privately maintained and authenticated broker, e.g. for enterprise applications. This figure shows several uses of tunnels:

Simple LoWPAN: Here, the LoWPAN Edge Router configures a tunnel to a known tunnel broker. From this tunnel it receives a global IPv6 prefix. Using this prefix it can autoconfigure its backhaul interface, and assign a prefix for its 6LoWPAN interface.

Extended LoWPAN: An IPv6 router in the local domain, in this case on the backbone link, configures a tunnel to a known tunnel broker. The router receives an IPv6 prefix from the tunnel, which it uses to configure its Internet interface, then to delegate a prefix to its backbone interface. This backbone prefix is then used by the edge routers on the backbone link, who assign the same prefix to their 6LoWPAN interfaces.

IPv4 remote server: Hosts with only IPv4 Internet access can also use a tunnel to enable IPv6 networking. In this case an IPv4 remote server can use a tunnel to receive IPv6 access, which it then uses to communicate with nodes in either 6LoWPAN network.

Note that these tunnels do not require use of the same tunnel end-point, as IPv6 packets are normally routed across the IPv6 Internet. Tunneling is a very useful technique which is applicable in both its forms, 6to4 automatic and 6in4 configured, to all kinds of 6LoWPAN networks.

5

Application Protocols

The Internet, and especially the Web, has become ubiquitous partly because of its ability to represent content in a universal way using a common application protocol – the *hypertext transfer protocol* (HTTP) [RFC2616]. While HTTP has become by far the most widely used application protocol, supporting web pages and web services, there are a large number of important application protocols used on the Internet. These include e.g. the file transfer protocol (FTP), real-time protocol (RTP), session initiation protocol (SIP), service location protocol (SLP) and the simple network management protocol (SNMP). These and other application protocols are crucial to the scale and breadth of the Internet as we know it today.

Application protocols can be defined as all the messages and methods having to do with inter-process communication via the Internet Protocol, as shown in Figure 1.8. The application layer depends on the transport layer to provide host-to-host communication and port multiplexing allowing multiple processes to communicate between end-points simultaneously.

The Internet of Things makes use of most of the same application protocols used in the Internet for communication between machines and services, for autoconfiguration, and for managing nodes and networks. Application protocols are equally important for the Wireless Embedded Internet with 6LoWPAN. However, 6LoWPAN is challenging in this respect. The limitations of 6LoWPAN such as small frame sizes, limited data rates, limited memory, sleeping node cycles, along with the mobility of devices make the design of new application protocols and the adaptation of existing ones difficult. Furthermore, the autonomous nature of simple embedded devices makes autoconfiguration, security and manageability all the more important.

In enterprise systems (a main application area of 6LoWPAN) the use of *web services* has become nearly ubiquitous over the past decade. Web services enable the communication between processes using well-defined message sequences with the *simple object access protocol* (SOAP), or stateless resources with *representational state transfer* (REST) style design. This allows business logic in distributed applications to exchange data, for machines to report measurements and for the remote management of devices. In order to integrate LoWPANs with enterprise systems that today run on web services, there is a great urge to

6LoWPAN: The Wireless Embedded Internet   Zach Shelby and Carsten Bormann
© 2009 John Wiley & Sons, Ltd

enable the use of web-service-related protocols between 6LoWPAN devices and services, as well as between 6LoWPAN devices.

This chapter introduces application design issues and protocols related to 6LoWPAN. Section 5.2 looks at design issues regarding 6LoWPAN, compression and security. Common protocol paradigms such as real-time, end-to-end and web services along with their applicability to 6LoWPAN are considered in Section 5.3. Finally a subset of interesting protocols usable with 6LoWPAN are introduced in Section 5.4. Protocols covered include web service protocols, MQTT, ZigBee CAP, service discovery protocols, SNMP, RTP and SIP. In addition, industry-specific protocols are considered in Section 5.4.7.

5.1   Introduction

Wireless Embedded Internet systems are usually designed for a specific purpose, for example a facility management network as described in Section 1.1.5 or for a simple home automation system. These two examples happen to have widely different application protocol requirements. Currently, large building automation systems are pre-configured to function in that environment, require management with e.g. SNMP, and often make use of industry-specific protocols such as BACnet. A home automation system on the other hand requires service discovery protocols such as SLP, and may make use of web-service style or proprietary protocols for data and management. What makes 6LoWPAN very different from vertical communication solutions is that the same network can be used by a large variety of devices running different applications thanks to the Internet model. All the protocols mentioned above can be run over the same IP network infrastructure, simultaneously. IP uses what is often called a *horizontal* networking approach.

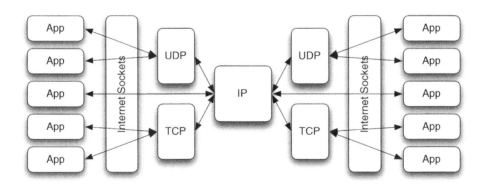

Figure 5.1 Applications process communication occurs through Internet sockets.

Although the Internet Protocol provides basic packet networking over heterogeneous links, it is UDP and TCP that allow for the large range of application protocols by providing *best-effort* (UDP) [RFC0768] and *reliable connection-oriented* (TCP) [RFC0793] multiplexed communications between application processes. IP protocols use a socket-based approach, where process *end-points* are identified by 16-bit source and destination port identifiers [RFC1122]. These are commonly called Internet sockets or network sockets. The concept is illustrated in Figure 5.1. The communication between any two end-points is uniquely identified for each transport by a four-tuple consisting of the local and remote socket addresses:

{src IP address, src port, dst IP address, dst port}

Application protocols use a socket API to access *datagram socket* (UDP) and *stream socket* (TCP) transport services along with *raw socket* (IP) services within a protocol stack. The different types of sockets are completely independent of each other (e.g. UDP port 80 and TCP port 80 can be used simultaneously). 6LoWPAN supports the compression of UDP ports down to a range of 16 [RFC4944], which is useful because a LoWPAN usually has a limited number of applications. Socket API programming for 6LoWPAN stacks is covered in Section 6.3. For a complete reference on Internet socket programming see [Stevens03].

TCP is not easy to compress, and is poorly suited to lossy wireless mesh networks because of its congestion avoidance design. For these reasons UDP is mainly used with 6LoWPAN as it is simple, compressible, and suits most applications protocol needs.

Figure 5.2 shows a layered map of all the protocols discussed in this chapter. This is by no means exhaustive as there are hundreds of IP-based protocols. There are however, a limited number which are suitable for use with 6LoWPAN. In the figure, these protocols are indicated in bold to indicate that they have either been designed for or are easily adaptable to 6LoWPAN. The MQ telemetry transport (MQTT) was developed by IBM for large-scale enterprise telemetry systems and is also suitable for use in sensor networks with MQTT-S. The ZigBee compact application protocol (CAP) allows for any ZigBee profile to be used over UDP, bringing ZigBee and IP closer together. Industry-specific protocols such as BACnet and oBIX for building automation are also covered. The simple location protocol (SLP) allows for service discovery while the simple network management protocol (SNMP) is widely used for management. Finally the real-time protocol allows for streamed real-time media to be transported over UDP which is an important function for audio, video and sensor data streams. In Section 5.4 these and other related protocols are discussed in detail.

Other widely used protocols such as TCP, HTTP, FTP, SIP and SOAP are depicted in Figure 5.2 as their importance on the Internet as a whole make them interesting candidates for use with 6LoWPAN, although considerable work still needs to be done to adapt or improve them. Examples of such adaptation efforts include e.g. embedded web services and TinySIP.

When designing application protocols for use with 6LoWPAN there are a number of requirements that need to be met. These are mostly due to the low-power, lossy nature of wireless mesh technologies along with limited frame sizes, limited memory in nodes, low data-rates and network simplifications. Furthermore, the nature of the embedded applications and battery-powered devices put new requirements on application protocols. The next section discusses 6LoWPAN application protocol design issues in detail.

5.2 Design Issues

Application protocols used over 6LoWPAN need to take a number of requirements into account which are typically not an issue over general IP networks. These issues include:

Link layer: Link-layer issues include lossy asymmetrical links, typical payload sizes of 70–100 bytes, limited bandwidth, and no native multicast support.

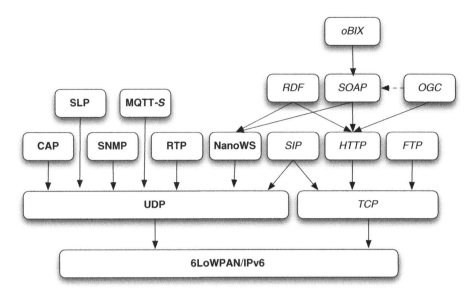

Figure 5.2 The relationship of common IP protocols.

Networking: Networking related issues include the use of UDP, limited compressed UDP port space and performance issues regarding the use of fragmentation.

Host issues: Unlike typical Internet hosts, 6LoWPAN hosts and networks are often mobile in nature during operation. Furthermore, battery-powered nodes use sleep periods with duty cycles often between 1–5 percent. A node may be identified in many ways, e.g. using its EUI-64, its IPv6 address or by a domain name, which should be taken into account.

Compression: The small payload sizes available often require compression to be used on existing protocols. Issues to consider include header and payload compression, and whether it is performed end-to-end or by an intermediate proxy.

Security: 6LoWPAN makes use of link-layer encryption which protects a single hop. Intermediate nodes are susceptible to attack, requiring sensitive application to employ end-to-end application level security. Edge routers need to implement firewalls in order to control the flow of application protocols in and out of LoWPANs.

Figure 5.3 illustrates where these issues typically occur in a LoWPAN. Mobility, node identification and sleep cycles are caused by node design and network properties. Intermediate 6LoWPAN Routers are a security risk, motivating end-to-end application security. The wireless link layer introduces bandwidth and frame size limitations. Finally, at the edge router we need to deal with compression, firewalls and UDP port space.

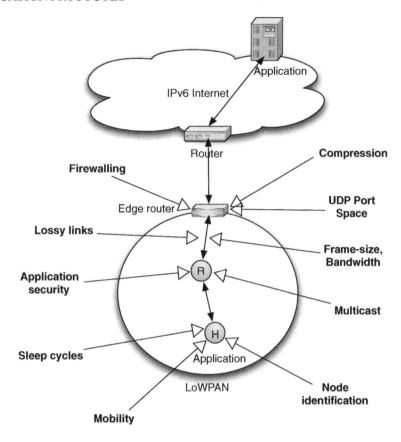

Figure 5.3 Application design issues to consider and where they occur in a LoWPAN.

5.2.1   Link layer

6LoWPAN enables the use of low-power radio technologies such as IEEE 802.15.4 and other ISM band radios. These radios are very different in nature from IEEE 802.11 WLANs (WiFi), Bluetooth or cellular radios, which support the use of standard IP protocols and applications. 6LoWPAN link-layer interaction was covered in Section 2.2.

Medium access control in e.g. IEEE 802.15.4 is achieved using carrier sense multiple access (CSMA) with a limited number of retransmissions for link-layer unicast frames [IEEE802.15.4]. In the presence of radio interference or packet collision there can be high packet loss ratios. Additionally the nature of radio propagation, heterogeneous transmission amplification and receiver sensitivity results in asymmetrical links, with packets successful in one direction but not in the other. Furthermore, radio fading and mobility causes the subset of neighbors within symmetrical range to vary from packet to packet. This must also be taken into account at the application layer, which should not assume too much regarding the stability of links. The lossy nature of links and the use of UDP motivates the use of end-to-end application reliability features.

Simple ISM band radios such as IEEE 802.15.4 rarely have support for native multicast. Instead they provide simple best-effort link-layer broadcast (MAC address 0xFFFF in IEEE 802.15.4). In a multihop network, multicast with a scope larger than link-local, is often mapped to a flood.

The most limiting feature of ISM band radios is their small frame size and limited bandwidth. IEEE 802.15.4 has a physical layer payload size 127 bytes in length, resulting in 72–116 bytes of available UDP payload depending on the MAC and 6LoWPAN features in use. Some link layers have even smaller frame sizes, whereas others may have frames as large as hundreds of bytes. The data rates over these radios are typically 20–250 kbit/s, shared by all nodes on the channel, and quickly reduced over multiple hops. Application protocols designed for use with 6LoWPAN should favor compact binary header and payload formats. Existing protocols should be optimized to reduce the size of packet payloads, compressed end-to-end, or compressed intermediately in order to maintain compatibility with existing Internet applications. Section 5.3.1 discusses end-to-end issues further.

5.2.2   Networking

6LoWPAN networking has features that place special requirements on application protocol design, although in general 6LoWPAN enables and simplifies the use of IPv6 over demanding link layers. Networking issues include the use of UDP, the compression of UDP ports and 6LoWPAN fragmentation.

As explained above, UDP has the most favorable characteristics for use over 6LoWPAN and is universally supported in protocol stacks. Although TCP has some justified use, it would require a new reliable transport or modified TCP to become universal over 6LoWPAN. Many Internet protocols today rely on TCP for a reliable connection-oriented byte stream. Instead, 6LoWPAN compatible application protocols mainly make use of UDP, which means the application protocol needs to deal with reliability if needed, out-of-order packets and datagrams rather than streams. If the UDP source or destination ports are compressed (as specified in [RFC4944] or [ID-6lowpan-hc]) then the port space can be limited down to 16 ports (ports 61616–61631). Although 6LoWPAN supports fragmentation in order to handle larger payloads coming in from outside the LoWPAN, the fragmentation of large payloads increases delay, packet loss probability and congestion. It is recommended to use application layer payload lengths that avoid the need for 6LoWPAN fragmentation whenever possible, as discussed in the previous section. Section 2.7 covered 6LoWPAN fragmentation in detail, including a discussion on performance implications.

5.2.3   Host issues

The *identification* of a device is especially important in embedded applications, for example in the monitoring of machines for maintenance. Typically a device can be identified by an application using some unique identifier such as its EUI-64, a serial number, the IPv6 address of the node, or by its domain name. The use of an IPv6 address for identifying 6LoWPAN devices is not recommended. The IPv6 address changes each time the LoWPAN Node or the whole LoWPAN changes its point of attachment, unless one of the node or network mobility solutions discussed in Section 4.1 is employed. A unique serial number such as the EUI-64 of the device is a reliable identifier, but must still be resolved to the IPv6 address of the device for communication by the application. The most application-friendly method is to use a domain name to identify a device, which is updated with the current IPv6 address of the device each time it moves, using appropriate DNS techniques. Section 4.1.3 discusses application methods for dealing with mobility.

The mobility of LoWPAN Nodes and networks causes further problems as nodes will often not be continuously available during handover between points of attachment. Battery-powered nodes are implemented to take advantage of aggressive sleep schedules in order to extend battery life. It is even common for a node to be active less than 1 percent of the time. The intermittent node availability due to mobility and sleep schedules needs to be taken into account during application design. For example, the synchronous polling of LoWPAN Nodes from a server should be avoided. Instead communication should be node initiated and asynchronous when possible.

5.2.4 Compression

As discussed above, the minimal payload available combined with fragmentation performance issues requires application protocols to use very compact formats. Most existing protocols were designed with other requirements in mind, such as human readability and extensibility, without payload size being an issue. Some application protocols are directly useful with 6LoWPAN such as RTP. Other existing protocols can or have been slightly adapted to make them efficient when used with 6LoWPAN; these include MQTT, SNMP, SLP and BACnet.

Application protocols designed for the Web, usually based on HTTP/TCP, are not well-suited for use over 6LoWPAN. HTTP uses a text-based human-readable format which takes space and is difficult to parse on simple embedded devices. XML is almost universally used for the machine-to-machine content carried in HTTP, such as SOAP. Although XML is a very useful, extensible markup on the Internet, it is much too sparse and complicated to parse for use with 6LoWPAN. Techniques to compress XML such as *binary XML* (BXML) and *efficient XML interchange* (EXI) along with *embedded web service* paradigms, are discussed in Section 5.4.1. When applying compression for web services, an important design consideration is whether to use compression *end-to-end* or to implement it with a *proxy*, e.g. on the edge router to the LoWPAN. End-to-end aspects are discussed more in Section 5.3.1.

5.2.5 Security

6LoWPAN depends on link-layer encryption for securing links in the LoWPAN. IEEE 802.15.4 includes a built-in 128-bit AES encryption feature which secures each link along the way. Link-layer encryption is, however, vulnerable at intermediate hops and understood by all nodes using the same encryption key. Therefore link-layer encryption is not very useful for securing application-level information which would be vulnerable at intermediate nodes, and over other IP networks after being routed outside the LoWPAN. Security issues were discussed in detail in Section 3.3.

If an application is working with sensitive data, then it should apply *end-to-end* application layer security. Here the application itself encrypts payloads at the sender of the packet, and decrypts them at the intended recipient. Many embedded enterprise systems may also deal with sensitive patient or customer information. As an added defense, application protocols on embedded devices may avoid sending personal identification information along with data to protect privacy. This matching would instead be performed in a back-end system.

Unlike personal computers, on which people make a large effort to maintain sophisticated firewalls for Internet security – embedded 6LoWPAN devices do not have the capabilities for complicated firewalls and are autonomous. Unrestricted data coming into a LoWPAN from the Internet can also easily overload the wireless network, causing denial of service. Special attention should be paid to firewall technology on the edge routers to LoWPANs to prevent unwanted application protocol traffic from entering and exiting LoWPANs, also avoiding denial of service situations. Application protocol designs need to take firewalls into account.

5.3   Protocol Paradigms

There is a basic set of paradigms by which most Internet application protocols function. These include the end-to-end paradigm, streaming, sessions, publish/subscribe and finally web services. In this section, we examine these paradigms in more detail along with their applicability to the Wireless Embedded Internet and their use in the protocols introduced in the next section.

5.3.1   End-to-end

The Internet socket model is based on the use of the underlying transport layer to provide a transparent datagram or byte stream service between application processes, or so-called application end-points. When considering the application layer this can be called an *end-to-end* paradigm where only the end-points participate in the application protocol exchanges. Some application protocols also include the possibility for intermediate nodes to inspect, cache or modify application protocols. Here we refer to this as *proxying*. An example would be an HTTP proxy that performs web-page caching. Figure 5.4 shows the difference between an end-to-end application protocol exchange, and a proxied one.

In the 6LoWPAN context, the end-to-end paradigm is significant in the realization of protocol compression, and in some cases to deal with the intermittent availability of mobile or battery-powered LoWPAN Nodes. Protocol compression of an existing protocol can be achieved either by supporting the compressed format natively on the IP application end-point, which is an end-to-end approach – or by having an intermediate proxy perform transparent compression so that IP applications do not need modification. A typical place to situate such a proxy would be on the edge router of a LoWPAN or on some local proxy server.

5.3.2   Real-time streaming and sessions

Many applications for embedded networks deal with *real-time* data streams, such as sensor data, audio or video. The Internet Protocol works on a best-effort approach, without *quality of service* (QoS) guarantees. Packets may arrive out-of-order or with significant jitter. 6LoWPAN applications performing real-time streaming need to take this into account. Typically UDP is employed for real-time applications, as a reliable transport like TCP may make jitter worse. Often it is better to drop a packet than to delay a real-time stream. Operations to be performed by an application protocol performing streaming include session

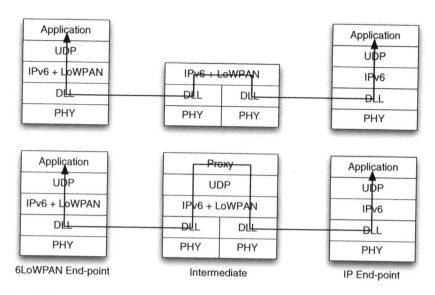

Figure 5.4 End-to-end (above) and proxied (below) application protocol paradigms.

setup (who participates in the stream), stream encoding, payload transmission and stream control.

Internet protocols already provide a good framework for working with real-time streams, which can be employed by 6LoWPAN applications as well. The real-time transport protocol (RTP) [RFC3550] encapsulates streams with appropriate timestamp and sequence information, while the companion RTP control protocol (RTCP) is used to control the stream. If a relationship between the sender(s) and receiver(s) of a stream needs to be automatically setup and configured, the session initiation protocol (SIP) [RFC3261] can be employed. These protocols are covered further in Section 5.4.6.

5.3.3   Publish/subscribe

Publish/subscribe (also known as pub/sub) is an asynchronous messaging paradigm in which publishers send data without knowing who the receiver is, and receivers subscribe to data based on the *topic* or content of the data. Pub/sub can be implemented using centralized *brokers* that match publishers and subscribers in a store-and-forward fashion, or in a distributed manner where subscribers filter messages directly from publishers. This decoupling of the application end-points allows for scalability and flexibility.

For the Internet of Things, pub/sub plays an important role as most applications are *data-centric*, i.e. it is not so important who sends data, but rather what the data is. One good example of a pub/sub protocol is the MQ telemetric transport (MQTT), which is a broker-based enterprise pub/sub protocol for telemetry, used widely by IBM [MQTT]. This has been adapted for use in sensor networks with MQTT-S, which is covered in Section 5.4.2 [MQTT-S].

5.3.4   Web service paradigms

Web services are defined by the W3C as a software system designed to support interoperable machine-to-machine communications over a network [WS]. Web services as a whole commonly work between clients and servers over HTTP. There are two different forms of web services: *service-based* (SOAP) web services and *resource-based* (REST) web services. Both forms of web services will play an important roll in 6LoWPAN applications, as discussed in the next section.

Service-based web services use XML following the SOAP format to provide *remote procedure-calls* (RPCs) between clients and servers [SOAP]. These SOAP messages and sequences can be described using the web services description language (WSDL) [WSDL]. This paradigm is widely used in enterprise machine-to-machine systems. A SOAP interface is typically designed with a single URL that implements several RPCs called *methods* (a good analogy is a verb) as in the following example:

```
http://sensor10.example.com/soap

Methods:
getSensorState(SensorID)
getSensorValue(SensorID)
setConfig(Parameter, Value)
getConfig(Parameter)
```

The representational state transfer (REST) paradigm instead models objects as HTTP *resources* (a good analogy is a noun), each with a URL accessible using standard HTTP methods [REST]. These interfaces can be described using the web application description language (WADL). With the release of WSDL 2.0, REST-based interfaces can alternatively be defined in a similar way to SOAP interfaces. This REST paradigm is widely used on the Internet between web sites. The content of REST HTTP messages can be of any MIME content, although XML is common in machine-to-machine applications. An example of a REST design follows, where objects are accessible using standard HTTP GET, POST, PUT and DELETE methods. In this example GET would be used on all resources to request the value, and POST would be used to set a new value for a parameter:

```
http://sensor10.example.com/sensors/temp
http://sensor10.example.com/sensors/light
http://sensor10.example.com/sensors/acc-x
http://sensor10.example.com/sensors/acc-y
http://sensor10.example.com/sensors/acc-z
http://sensor10.example.com/config/sleeptime
http://sensor10.example.com/config/waketime
http://sensor10.example.com/config/enabled
http://sensor10.example.com/config/samplerate
```

5.4   Common Protocols

This section introduces protocols that are commonly used or have good potential for use over 6LoWPAN. These include web service protocols, MQTT-S, ZigBee CAP, service discovery protocols, SNMP, RTP/RTCP, SIP and industry-specific protocols.

5.4.1 Web service protocols

The web service concept is hugely successful on the Internet, especially in enterprise machine-to-machine Internet systems. As many back-end systems incorporating information from LoWPAN devices will already be using existing web service principles and protocols, it is expected that 6LoWPAN will be integrated into the web service architecture. The use of XML, HTTP and TCP makes the adaptation of web services challenging for LoWPAN Nodes and networks. In this section we look inside web service protocols, and the technologies that can adapt them for use with 6LoWPAN.

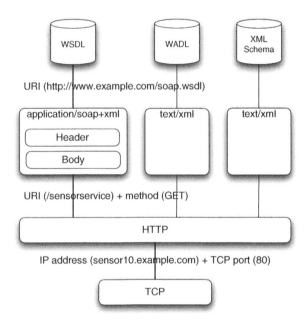

Figure 5.5 Typical structure of web service content over HTTP/TCP.

Figure 5.5 shows the typical structure of web service content, which is always built upon HTTP and TCP as used today on the Internet. Web services are simply URLs available on an HTTP server with services or resources accessible behind them. In the SOAP model a URL identifies a service, e.g. /sensorservice in the figure. This service may support any number of methods (e.g. GetSensor) with corresponding responses that are described by a WSDL document. SOAP (application/soap+xml) is an XML format consisting of a header and a body, in which the body carries any number of messages.

Resource-based web services can also be realized using a REST design. Figure 5.5 also shows text/xml content used directly over HTTP. In this model formal message sequences are not used, instead each resource is identified by a URL. By using different HTTP methods on that URL, the resource can be accessed. For example sending an HTTP GET for /sensors/temp might return a text/xml body with the temperature of the sensor. REST designs make use of well-known XML or other formats to give meaning to the content that can be understood by all parties.

All web services have the same basic problems for 6LoWPAN use. XML is typically too large for marking up content in the payload space available, HTTP headers have high overhead and are difficult to parse, and finally TCP has limitations of its own. A simplified HTTP header with application/soap+xml content for the example above may look like:

```
POST /sensorservice HTTP/1.1
Host: sensor10.example.com
Content-Type: application/soap+xml; charset=utf-8
Content-Length: nnn

<?xml version="1.0"?>
<soap:Envelope
xmlns:soap="http://www.w3.org/2001/12/soap-envelope"
soap:encodingStyle="http://www.w3.org/2001/12/soap-encoding">

<soap:Body xmlns:m="http://www.example.com/soap.wsdl">
  <m:GetSensor>
    <m:SensorID>0x1a</m:SensorID>
  </m:GetSensor>
</soap:Body>

</soap:Envelope>
```

This simple example has a length of 424 bytes, which may require up to six fragments to transmit it over a 6LoWPAN network. Next we look at different technologies to allow web services to be used with 6LoWPAN. There are two fundamental ways to integrate 6LoWPAN into a web service architecture: using a *gateway approach* or a *compression approach*.

Gateway approach

In the gateway approach, a web service gateway is implemented at the edge of the LoWPAN, often on a local server or the edge router. Inside the LoWPAN a proprietary protocol is used to request data, perform configuration etc. The gateway then makes the content and control of the devices available through a web service interface. In this approach web services actually end at the gateway. This technique is widely used with non-IP wireless embedded networks such as ZigBee and vendor-specific solutions. One downside of this approach is that a proprietary or 6LoWPAN-specific protocol is needed between nodes and the gateway. Furthermore, the gateway is dependent on the *content* of the application protocols. This creates scalability and evolvability problems where each time a new use of the LoWPAN is added or the application format is modified, all gateways need to be upgraded.

Compression approach

In the compression approach, the web service format and protocols are compressed to a size suitable for use over 6LoWPAN. This can be achieved using standards, and has two forms: end-to-end and proxy. In the end-to-end approach the compressed format is supported by both application end-points. In the proxy approach an intermediate node performs transparent compression so that the Internet end-point can use standard web services.

Several technologies exist for performing XML compression. The WAP Binary XML (WBXML) format was developed for mobile phone browsers [WBXML]. Binary XML (BXML) from the Open Geospatial Consortium (OGC) was designed to compress large sets of geospatial data and is currently a draft proposal [BXML]. General compression schemes like Fast Infoset (ISO/IEC 24824-1) work like zip for XML [FI]. Finally, the W3C is currently completing standardization of the efficient XML interchange (EXI) format, which performs compact binary encoding of XML [EXI]. One suitable technology for use with 6LoWPAN is the proposed EXI standard, as it supports out-of-band schema knowledge with a sufficiently compact representation. LoWPAN Nodes do not actually perform compression; instead they directly make use of the binary encoding for content, which keeps node complexity low.

XML compression alone only solves part of the problem. HTTP and TCP are still not suitable for use over 6LoWPAN. One commercial protocol solution called Nano Web Services (NanoWS) from Sensinode applies XML compression in an efficient binary transfer protocol over UDP, which has been specifically designed for 6LoWPAN use [Sensinode]. The SENSEI project is also researching the use of embedded web services inside Internet-based sensor networks [SENSEI]. The ideal long-term solution will be the standardization of a combination of XML binary encoding bound to a suitable UDP-based protocol.

The *namespace* and *schema* used with 6LoWPAN devices must also be carefully designed. Standard schemas such as SensorML from the OGC [SensorML] are often much too large and complicated for efficient use over 6LoWPAN even with compression. The most efficient result is achieved using a simple schema or one designed specifically for use with embedded devices.

5.4.2   MQ telemetry transport for sensor networks (MQTT-S)

The MQ telemetry transport (MQTT) is a lightweight publish/subscribe protocol designed for use in enterprise applications over low-bandwidth wide area network (WAN) links such as ISDN or GSM [MQTT]. The protocol was designed by IBM and is used in commercial products such as Websphere and Lotus, enjoying widespread use in M2M applications. MQTT uses a broker-based pub/sub architecture, to which clients publish data based on matching topic names. Subscribers then request data from the broker based on topic names. Although MQTT was designed to be lightweight, it requires the use of TCP, and the format is inefficient over 6LoWPAN networks.

In order to allow for MQTT to be used also in sensor networks, MQ telemetry transport for sensor networks (MQTT-S) was developed [MQTT-S]. This optimized protocol can be used over ZigBee, UDP/6LoWPAN or any other simple network providing a bi-directional datagram service. MQTT-S is optimized for low-bandwidth wireless networks with small frame sizes and simple devices. It is still compatible with MQTT and can be seamlessly integrated with MQTT brokers using what is called an MQTT-S gateway.

The MQTT-S architecture is shown in Figure 5.6. It is made up of four different elements: MQTT brokers, MQTT-S gateways, MQTT-S forwarders and MQTT-S clients. Clients connect themselves to a broker through a gateway using the MQTT-S protocol. The gateway may be located e.g. on the LoWPAN Edge Router, or it may be integrated in the broker itself, in which case MQTT-S messages are simply carried end-to-end over UDP. Gateways translate between MQTT-S and MQTT. In case a gateway is not directly available, forwarders are used to forward (unmodified) messages between clients and brokers. Forwarders may not be needed with 6LoWPAN as UDP datagrams can be sent directly to a gateway.

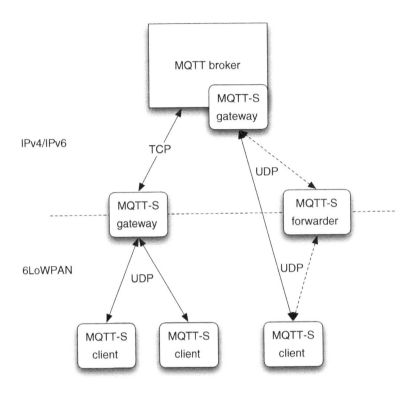

Figure 5.6 The MQTT-S architecture used over 6LoWPAN.

Figure 5.7 shows the MQTT-S message structure, which consists of a length field, a message type field and then a variable-length message part. Next the basic functionality of MQTT-S is described shortly. Readers should consult the MQTT-S specification for full protocol details [MQTT-S].

Figure 5.7 The MQTT-S message structure.

Protocol operation

MQTT-S includes a gateway discovery procedure, which does not exist in MQTT. Alternatively clients can be pre-configured with the gateway location to avoid discovery overhead. Gateways send periodic ADVERTISE messages, and clients may send SEARCHGW messages. A GWINFO message is sent to a client in response to SEARCHGW with basic information about the gateway.

Clients CONNECT to a gateway which responds with an ACK. DISCONNECT is used to end a connection or to indicate a sleep period. Clients can connect with multiple gateways (and thus brokers) which are able to perform load balancing. Gateways can function in either *transparent mode*, where a connection to the broker is maintained for each client, or in *aggregation mode*, where the gateway aggregates messages from all clients into a single broker connection. Aggregation mode can considerably improve scalability.

MQTT-S makes use of two-byte *topic IDs* and short *topic names* to optimize the long topic name strings normally used in MQTT. MQTT-S includes a registration message REGISTER, which explicitly indicates the topics that a client is publishing. This reduces the bandwidth compared to MQTT where the topics and data are published in the same message. A client can send a REGISTER message with the topic name, which is acknowledged with a REGACK indicating the assigned topic ID. Topic names and IDs can also be pre-configured to avoid the need for registrations in certain cases. The client then publishes its data with PUBLISH messages including the topic ID and possible QoS information.

Clients subscribe by sending SUBSCRIBE to the gateway including the topic name of interest, which is acknowledged with SUBACK including an assigned topic ID. UNSUBSCRIBE is used to remove a subscription from the gateway.

5.4.3   ZigBee compact application protocol (CAP)

The ZigBee application layer (ZAL) [ZigBee], and the ZigBee cluster library (ZCL) [ZigBeeCL] specify an application protocol enabling interoperability between ZigBee devices at the application layer. The ZigBee Alliance maintains a series of specifications for ad hoc networking between embedded devices using a single radio, IEEE 802.15.4. Typical applications for ZigBee include home automation, energy applications and similar local-area wireless control applications. ZigBee makes use of a vertical *profile* approach over the ZAL and ZCL, with profiles for different industry applications such as the ZigBee home automation profile [ZigBeeHA] or the ZigBee smart energy profile [ZigBeeSE]. The ZAL and ZCL provide the key application protocol functionality in ZigBee, enabling the exchange of commands and data, service discovery, binding and security along with profile support. These protocols use compact binary formats with the goal of fitting in small IEEE 802.15.4 frames.

The ZigBee application protocol solution would have benefits used over standard UDP/IP communications as well, especially over 6LoWPAN, as the ZigBee application protocol has been designed with similar requirements. This would allow much wider use of ZigBee profiles, also end-to-end over the Internet eliminating the need for gateways. However the ZAL had been originally designed with only the ZigBee network layer primitives and IEEE 802.15.4 in mind.

A solution for using ZigBee application protocols and profiles over UDP/IP has been proposed in [ID-tolle-cap], which is an IETF Internet draft. This specification defines how the

ZAL is mapped to standard UDP/IP primitives, enabling the use of any ZigBee profile over 6LoWPAN or standard IP stacks. This adaptation of the ZAL for use with UDP/IP is called the *compact application protocol* (CAP). The CAP protocol stack is shown in Figure 5.8. The functions of the ZAL and ZCL are implemented by the CAP. The data protocol corresponds to the ZigBee cluster library. The management protocol corresponds to the ZigBee device profile handling binding and discovery. Finally the security protocol implements ZigBee application sublayer (APS) security. Any ZigBee public or private application profile can be implemented over CAP in the same way that it would use the native ZigBee ZAL/ZCL. This allows for ZigBee application profiles to directly be applied to IP networks.

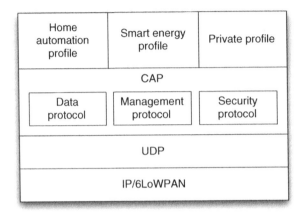

Figure 5.8 The CAP protocol stack.

The main modification to the ZAL has to do with using IP hosts and IP addresses instead of IEEE 802.15.4 hosts and IEEE 802.15.4 addresses. ZigBee application layer messages are placed inside UDP datagrams using the CAP data protocol instead of ZigBee network layer frames. To receive unsolicited notifications CAP listens to a well-known UDP port. The ZAL identifies nodes by their 64-bit or 16-bit IEEE 802.15.4 MAC address. In CAP this is replaced by a CAP address record which can contain an IPv4 address plus UDP port, IPv6 address plus UDP port, or a fully qualified domain name plus UDP port.

It is assumed that the CAP is configured with the IP address and port of the discovery cache, trust center, binding coordinator and binding cache during bootstrapping. These can be configured manually, using DHCP, a special DNS entry or using a CAP server discovery message.

The CAP protocol is simply ZigBee application layer APS frames placed in UDP, with all the standard options and extensions. The APS delivery modes are mapped to IP unicast and broadcast delivery, and groupcast is reduced down to broadcast. CAP supports secure transmission and the use of APS acknowledgments, which provide limited application protocol reliability. The CAP data protocol is contained within the APS payload and it contains the ZCL command frame, with support for all ZCL command types. None of the ZCL commands require modification for use with CAP. The CAP management protocol modifies the ZigBee device profile command frame to remove IEEE 802.15.4 specific frames or to modify the address where possible. The ZigBee security and key management features are implemented by the CAP security protocol. Security issues are considered in detail in [ID-tolle-cap].

This proposal [ID-tolle-cap] for using ZigBee application profiles, service discovery, binding and application-layer security over UDP/IP is an important step towards the convergence of ZigBee and the Wireless Embedded Internet. IP provides a wealth of advantages compared to ZigBee networking, including seamless Internet integration, better scalability, link-layer independence and applicability to a much larger range of applications. At the same time the ZigBee application layer and public profiles provide important features currently lacking in application protocols suitable for 6LoWPAN. The CAP is, however, a suitable solution only for a subset of Internet of Things applications such as home automation. It complements enterprise telemetry protocols such as MQTT-S, and general-purpose embedded web services. Thanks to 6LoWPAN's IP design and the socket concept, all these protocols can be used together over UDP/6LoWPAN networks. CAP is a concept that would need standardization to become widely used; potential bodies to do that include the IETF and the ZigBee Alliance. The upcoming ZigBee/IP Smart Energy 2.0 profile will not be based on a CAP approach. Instead, open IETF, W3C and IEC standards are being selected for application protocols and data formats to enable end-to-end use over the Internet.

5.4.4   Service discovery

Service discovery is an important issue in Wireless Embedded Internet applications, where devices are autonomic – also requiring the autoconfiguration of applications. Service discovery is used to find which services are offered, what application protocol settings they use, and at what IP address they are located. Typical protocols used for service discovery on embedded devices includes the service location protocol (SLP), universal plug-n-play (UPnP) and devices profile for web services (DPWS). Some application protocols such as ZigBee CAP or MQTT-S have their own built-in discovery features. Frameworks such as OGC or SENSEI also have built-in service discovery and description mechanisms.

The service location protocol is used for general service discovery over IP networks [RFC2608]. SLP needs optimizations in order to be effectively used with 6LoWPAN because of the size of typical messages. There has been a proposal for a simple service location protocol (SSLP) [ID-6lowpan-sslp], which provides a simple, lightweight protocol for service discovery in 6LoWPAN networks. Such a protocol could be easily interconnected with SLP running on IP networks by an SSLP translation agent located on an edge router – thus allowing 6LoWPAN services to be discovered from outside the LoWPAN and vice versa. SSLP supports most of the features of SLP, including the optional use of directory agents. The SSLP header format consists of a four-byte base header followed by specific message fields. As in SLP, service types, scopes and URLs are carried as strings. Strings used with a scheme like SSLP should be kept as short as possible.

UPnP is a protocol aimed at making home devices automatically recognizable and controllable as specified in [UPnP]. UPnP makes use of three protocols: the simple service discovery protocol (SSDP) for discovering devices, the generic event notification architecture (GENA) for event notification and SOAP for controlling devices. Devices descriptions are stored as XML and are retrieved using HTTP after initial discovery using SSDP. UPnP is not directly applicable to 6LoWPAN devices because of its dependence on broadcast along with XML- and HTTP-based descriptions and protocols. SSDP may be applicable directly over 6LoWPAN as it is similar to SSLP. It may be possible to use UPnP, or a subset of it, over 6LoWPAN with web service compression and binding applied to UPnP descriptions and protocols. However, this would require a special version of UPnP to be specified for use over 6LoWPAN and similar networks.

The devices profile for web service (DPWS) describes a basic set of functionality to enable embedded IP devices with web-service-based discovery, device descriptions, messaging and events [DPWS]. The objectives of DPWS are similar to those of UPnP, but DPWS uses a pure web-service approach. DPWS has recently been standardized under OASIS. As DPWS descriptions are XML and all messaging is based on XML/HTTP/TCP it would require web service compression and binding, along with simplification, in order to be used over 6LoWPAN. DPWS has been gaining in popularity for use in enterprise and industrial systems as devices using DPWS can be automatically integrated into back-end systems based on web services.

5.4.5   Simple network management protocol (SNMP)

Network management is an important feature of any network deployment, and a certain amount of management is necessary even for autonomous wireless embedded devices. There are several ways of performing management in IP networks, e.g. the simple network management protocol (SNMP), web services or proprietary protocols. SNMP is a standard for the management of the network infrastructure and devices in IP networks. It includes an application protocol, a database schema and data objects. The current version is SNMPv3, specified in [RFC3411]–[RFC3418]. SNMP exposes variables to a management system which can be GET or in some cases SET in order to configure or control a device. The variables exposed by SNMP are organized in hierarchies called management information bases (MIBs).

The polling approach used by SNMP (GET messages) is the biggest drawback of the approach. Polling approaches don't work for battery-powered LoWPAN Nodes which use sleep schedules, and blindly polling for statistics (which may not have changed) creates unnecessary overhead. An event-based approach would need to be added to SNMP for applicability to 6LoWPAN management.

The suitability of using SNMPv3 with 6LoWPAN has also been analyzed in [ID-snmp-optimizations], which found that optimizations are needed to reduce the packet size and memory cost. Furthermore a MIB has been specified for 6LoWPAN in [ID-6lowpan-mib]. The following optimizations for SNMPv3 have been identified in these drafts:

- Currently SNMPv3 requires the handling of payload sizes up to 484 bytes, which creates too much overhead for managing large 6LoWPANs.

- The SNMPv3 header is variable in size, and needs to be optimized for 6LoWPAN. Only a minimal subset of functionality should be supported and the header size should be limited.

- The binary encoding rules (BER) of the payload use variable length fields. For 6LoWPAN, fixed length fields or more compact encoding may be necessary.

- Payload compression and aggregation may be needed for 6LoWPAN.

- To reduce memory requirements the maximum size of SNMP messages should be limited. Furthermore MIB support should be carefully chosen.

5.4.6 Real-time transport and sessions

The real-time transport protocol (RTP) [RFC3550] is used for the end-to-end delivery of *real-time* data. RTP is designed to be IP-version- and transport-independent, and can be used over both UDP and TCP. The base RTP header provides basic features for end-to-end delivery: payload-type identification, a sequence number and a timestamp. The RTP header format is shown in Figure 5.9. RTP does not by itself provide any kind of QoS, but it is able to help deal with out-of-order packets and jitter with the sequence number and timestamp fields, respectively. The accompanying real-time control protocol (RTCP) is used during an RTP session to provide feedback on the QoS of RTP data delivery, to identify the RTP source, adjust the RTCP report interval and to carry session control information. RTP uses the concept of *profiles*, which define possible additional headers, features and payload formats for a particular class of application. The RTP audio video profile (AVP) [RFC3551] specifies the profile for common audio and video applications.

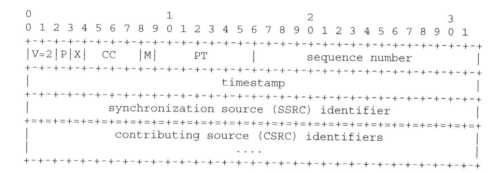

Figure 5.9 The RTP base header.

RTP is applicable also over 6LoWPAN for the transport of real-time data. As RTP makes use of UDP, is IP-version-independent and has a fairly compact header format, it is directly usable without modification. Existing profiles such as AVP are useful for the streaming of low-rate audio and video using the most efficient codecs, and custom profiles can be created for the transmission of e.g. sensor data streams.

Although RTP can be used to delivery and monitor real-time data streams, it requires that the sender and receiver somehow know about and find each other. Although this may be the case in specialized embedded applications of RTP over 6LoWPAN, the automatic negotiations of real-time sessions or other messaging may be very useful. The session initiation protocol (SIP) [RFC3261] was designed for establishing, modifying and tearing down multimedia sessions over IP. SIP is widely used for Voice-over-IP (VoIP) applications, and forms the backbone for the IP Multimedia System (IMS) upon which future cellular services will be built. The SIP design is similar to that of HTTP, in that it uses a human-readable header format. SIP can be used over either UDP or TCP, can be handled by intermediate proxies, and provides identifiers for dealing with mobility. SIP exchanges are typically performed between SIP user agents (e.g. embedded devices) and servers. Typical methods include REGISTER, INVITE, ACK and BYE. SIP uses a separate session description protocol (SDP) [RFC2327] to negotiate media types. An example SIP header for an INVITE from [RFC3261] is shown below:

```
INVITE sip:bob@biloxi.com SIP/2.0
Via: SIP/2.0/UDP pc33.atlanta.com;branch=z9hG4bKnashds8
Max-Forwards: 70
To: Bob <sip:bob@biloxi.com>
From: Alice <sip:alice@atlanta.com>;tag=1928301774
Call-ID: a84b4c76e66710
CSeq: 314159 INVITE
Contact: <sip:alice@pc33.atlanta.com>
Content-Type: application/sdp
Content-Length: 142
```

The SIP header and body format is typically too large for efficient use over 6LoWPAN. But as SIP can be applied to session setup, alarms, events and IMS integration, it has valuable use in low-power embedded networks. One solution for using SIP with sensor networks is TinySIP, which defined alternative messages for use with TinyOS networking that were then mapped to SIP by a gateway application [TinySIP]. As SIP is similar to HTTP, the web service compression and UDP binding techniques discussed in Section 5.4.1 may also be applicable through further development.

5.4.7   Industry-specific protocols

This section gives an overview of industry-specific application protocols that can be used over IP, and are relevant for Wireless Embedded Internet applications using 6LoWPAN. Building automation and energy are good examples of industries that have traditionally specified their own application protocols and formats. These are enterprise applications where system integrators make use of equipment from multiple vendors together with back-end computer systems to achieve large deployments. The need for common application protocols and formats is obvious in such an environment. As communication technology has evolved, industry-specific protocols have steadily evolved to enable use over IP. Many industry-specific protocols may be used over 6LoWPAN, whereas others may require the addition of compression, IPv6 support or UDP support for example. Next we examine some common building automation and energy industry application protocol standards.

BACnet

The building automation and control networks (BACnet) standard was created by the American Society of Heating Refrigeration and Air-Conditioning Engineers (ASHRAE) in 1995 to bring interoperability to HVAC building automation. BACnet is published as ANSI standard 135 [BACnet] and ISO 16484-5. Since its original release it has developed into a broad building automation standard used by over 350 vendors. The latest version was published in 2008.

BACnet is a network and application protocol format with support for a wide range of communication technologies including Ethernet, RS232, RS-485 and LonTalk. BACnet includes support for use over UDP/IP known as BACnet/IP. The standard BACnet network and application protocol frames are carried over UDP by encapsulating them in a BACnet virtual link layer (BVLL). This adaptation binds BACnet to an underlying communication technology. Currently there is only a BVLL defined for use with IPv4, but an extension of that for IPv6 is straightforward. IPv6 support is current under design in the BACnet IP working group. BACnet makes use of unicast, broadcast and optionally multicast IP communications, and is based on an object-oriented design. BACnet objects have properties that are acted upon using protocol services. These protocol services include Who-Is, I-Am, Who-Has and I-Have used for device and object discovery along with read-property and write-property used for data access.

As BACnet was designed for a whole range of low-bandwidth links, and has native support for IPv4, it will be a useful protocol over 6LoWPAN for building automation applications. The adaptation of BACnet/IP for use with 6LoWPAN will require IPv6 support, and should make careful use of multicast – keeping in mind its overhead on wireless multihop mesh networks. The performance and possible optimization of BACnet service protocol traffic over low-power wireless mesh networks should also be studied as BACnet currently assumes wired links.

KNX

Konnex (KNX) is an open protocol for home and building automation standardized internationally (ISO/IEC 14543-3), in Europe (CENELEC EN 50090) and in China (GB/Z 20965) [KNX]. It is also published as ANSI standard 135. KNX is based on the convergence of three previous European standards in the home and building automation domain. It is supported by over a 100 different manufacturers and promoted by the Konnex Association. It is estimated that over 80 percent of the home automation devices sold in Europe use KNX.

The KNX protocol supports several different communication media: twisted pair, power-line, radio frequency (RF) and IP. Twisted pair is the most common KNX medium and is typically installed when a building or house is constructed. Powerline and RF are often used when retrofitting existing buildings. The KNX RF specification uses its own framing over an 868 MHz radio at 16 kbit/s. KNX networks can theoretically support up to 64k devices using twisted pair, power-line or RF communications.

KNX has some support for IP, also known as KNXnet/IP. KNX IP support is part of a framework called ANubis (advanced network for unified building integration and services) and among many other things provides a way to encapsulate KNX frames over IP. The purpose of this is currently to interconnect KNX networks using IP networks, to enable remote monitoring and to interconnect with other systems such as BACnet. This IP encapsulation technique could also be usefully applied over low-power IP and 6LoWPAN networks to KNX devices themselves. This would need a considerable amount of development and standardization.

oBIX

The open building information exchange (oBIX) is a web-service-based standard for access to building control information [oBIX]. The standard is meant to provide high-level access using an open, universal format between back-end systems and building control networks, and to interconnect building control networks. oBIX is standardized by the Organization for the Advancement of Structured Information Standards (OASIS), and v1.0 was released in 2006.

oBIX provides a web service interface, which can be used to interact with any building automation network including BACnet, KNX, Modbus, Lontalk or proprietary networks using the oBIX XML format. The format provides normalized representation of constructs common to building automation protocols: points (scalar value, status), alarms and histories. It has an extensible meta-format which can be used to describe any system. oBIX provides a low-level object model for working with these constructs. Usually these are accessed by using generic oBIX constructs, for example by an enterprise developer.

oBIX is web service binding agnostic. It can be used over both SOAP and directly over HTTP in a REST style. It represents objects with URLs and object state with XML. oBIX may be applicable also for use directly in building automation networks thanks to 6LoWPAN. Instead of running a control network specific building automation protocol such as BACnet/IP or KNX over 6LoWPAN, oBIX together with compression and UDP/IP binding may be a solution. Careful design of the oBIX objects and elements used would be important to keep packet sizes reasonable.

ANSI C12.19

Smart metering, smart grids and automatic metering infrastructures (AMI) are developing very rapidly with the need for energy savings and the demand for energy. The American National Standards Institute (ANSI) C12 family of standards defines formats, interfaces and protocols for the utility industry, and is widely used in North America [ANSI]. In particular, one set of standards within ANSI C12 specifies communications for utility end devices (usually electric, gas and water meters) to communicate together and with back-end systems. These standards enable the configuration, programming and monitoring of these devices remotely over both point-to-point and IP networks.

The ANSI C12.18 standard defines a point-to-point optical interface along with the protocol specification for electric metering (PSEM). The ANSI C12.19 standard defines utility industry end device data tables, which are accessible using PSEM. ANSI C12.21 specifies a communication protocol over modem lines. Finally a new standard ANSI C12.22 specified the interfacing of devices to any data communications network including Internet protocols. This furthermore specifies the extended protocol specification for advanced metering (EPSEM).

The ANSI C12 PSEM protocol and device data formats (tables) were designed with simple embedded electric meters and low-bandwidth point-to-point links in mind. Therefore PSEM uses a compact binary encoding for all its protocol and data fields. The typical frame size expected by point-to-point links is 64 bytes in the ANSI C12.22 specification. The ANSI C12.22 specification is therefore well suited for use over 6LoWPAN. The specification includes a simple example of using the application layers with a TCP/IPv4 communication stack. The standard may also be used with a UDP/IPv6 stack, as the protocol includes acknowledgment features for reliability along with application-layer segmentation and reassembly. Application layer security features are built in elements of the protocol.

DLMS/COSEM

The device language message specification (DLMS) is a European model of communication exchange used to interact with utility end devices for meter reading, tariff and load control. It uses the companion specification for energy metering (COSEM) as a data exchange format and protocol. Together, they are standardized by the IEC under the 62056 series [IEC62056]. The DLMS Association [DLMS] actively promotes the maintenance and use of the standards.

IEC 62056 is similar in function to its American counterpart, ANSI C12. It makes use of local optical or current loops to meters, point-to-point serial modems, or IP networks for communication. The COSEM protocol defines the application layer in IEC 62056-53. As with ANSI C12, an object model is used to access information as defined in IEC 62056-61. The use of COSEM over IPv4 is described in detail in IEC 62056-47. This specification describes transport using both UDP and TCP, which enables use over 6LoWPAN. Although the specification describes the use of IPv4, the changes needed for IPv6 are minor. It is unclear if any changes would be needed to the standard to allow the use of IPv6. Even though the data representation format is not as compact as in ANSI C12, it is still suitable for the frame size and bandwidth limitations of 6LoWPAN.

6

Using 6LoWPAN

This chapter gives an overview of implementation issues to consider when integrating 6LoWPAN in wireless embedded devices and routers. Integrating communications into an embedded device is a more complicated task than on a PC, which typically comes with a standard IP protocol stack, network interfaces and drivers by default. Embedded devices have tightly integrated hardware and software designs, often without the possibility for the complete hardware abstraction and generic interfaces found in PC architectures. In most applications for the Wireless Embedded Internet, devices are expected to be low cost, low power and compact. Embedded integration is maximized by the use of *system-on-a-chip* (SoC) radio technology where the transceiver, microcontroller, protocol stack and application are all integrated on a small, inexpensive chip.

The use of 6LoWPAN with embedded devices (such as the one shown in Figure 6.1), as with any other embedded communications, needs special design consideration. In this chapter we look at 6LoWPAN chip, stack integration, wireless node application development and edge router issues. An overview is also given of common open source and commercial protocol stacks for 6LoWPAN and ISA100.

Figure 6.1 An example embedded device using a modular two-chip (MSP430+CC2420) design.

The accompanying course material and exercises for this book are an excellent companion to this chapter and are available at http://6lowpan.net. Exercises and examples with the course material are provided for the uIPv6 stack on Contiki, which is introduced in this

6LoWPAN: The Wireless Embedded Internet   Zach Shelby and Carsten Bormann
© 2009 John Wiley & Sons, Ltd

chapter and in the exercises. We recommend that the reader actually installs Contiki [Contiki] (http://www.sics.se/contiki) along with the book exercises and gives the hands-on implementation of a 6LoWPAN application a try! The Instant Contiki package is a quick way to get started, and runs in a virtual machine under any operating system.

6.1 Chip Solutions

As 6LoWPAN is a networking technology, and is used in embedded devices, a 6LoWPAN protocol stack is embedded on a microcontroller somewhere in the device. There are three different models typically used to embed such a wireless protocol solution: single-chip, two-chip and network processor solutions. This section looks at the benefits and applicability of these different options. A *single-chip* solution uses a system-on-a-chip radio with a built-in microcontroller, whereas the *two-chip* solution uses a separate microcontroller together with a radio *transceiver*. *Network processor* solutions use a radio transceiver that includes the protocol stack, which is used by a separate *application processor*.

6.1.1 Single-chip solutions

In devices where minimizing cost and size is critical, while at the same time the complexity of the embedded application is low, single-chip solutions are attractive. In a single-chip solution SoC radio technology is used where the radio front-end, transceiver and microcontroller are integrated together with flash, memory and other peripherals. A very small bill-of-materials is usually required in addition to the SoC in order to make an independent wireless node, usually including RF matching components, the antenna, crystals and power supply, along with possible sensors and actuators. Figure 6.2 shows a block diagram of a single-chip architecture. All the software for the device runs on the SoC microcontroller and is stored in its flash. This includes hardware drivers, a possible embedded operating system (OS), the 6LoWPAN protocol stack and the device applications. Examples of SoC hardware solutions include the TI CC2530, TI CC1110 and the Jennic JN5139. Most 6LoWPAN protocol stacks can be integrated into single-chip solutions.

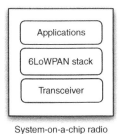

System-on-a-chip radio

Figure 6.2 Single-chip solution architecture.

The downside of this approach is that the integration of all software on the same microcontroller increases complexity and development time. As SoCs make use of small, specialized microcontrollers with no memory protection, the integration of applications with the protocol stack, underlying drivers and OS takes more effort as each configuration needs extensive testing. At the same time this approach limits the reusability of applications written for that SoC because of specialized compilers and stack solutions.

6.1.2 Two-chip solutions

In devices where a microcontroller architecture has already been chosen, or where application complexity and performance requirements are great, then two-chip solutions tend to be applied. Figure 6.3 shows a block diagram of the two-chip architecture. In a two-chip solution, the application processor and transceiver are separate. A variation of this approach, where the transceiver also includes the protocol stack, is called a network processor and is discussed in the next section. An application processor typically communicates with the transceiver over a universal asynchronous receiver/transmitter (UART) or serial peripheral interface (SPI). The 6LoWPAN protocol stack is integrated with the embedded application, drivers and OS on the application microcontroller. This leaves the developer almost complete freedom in the choice of application microcontroller, which may have special embedded control, signal processing or performance requirements. A two-chip solution can be implemented using any microcontroller with suitable flash and memory. Examples of common radio transceivers include the TI CC2520 and the Atmel AT86RF231.

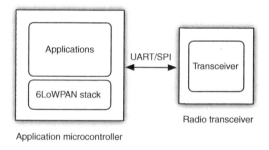

Figure 6.3 Two-chip solution architecture.

The downside of this approach is that the 6LoWPAN protocol stack and the embedded application need to be integrated in the same microcontroller. As with the single-chip solution, this integration may require extensive engineering and testing. In addition, many protocol stacks are tied to a particular microcontroller or radio transceiver, making it difficult to move between microcontrollers.

6.1.3 Network processor solutions

For projects in which a design or application software already exists, or for minimum engineering effort in new devices, a two-chip *network processor* solution is appropriate. As with the previous two-chip solution, the network processor is a separate chip. However, in this case a 6LoWPAN protocol stack is included with the transceiver. This architecture is shown in Figure 6.4. A network processor is often realized on an SoC radio chip, with network processor firmware. Typically a much smaller SoC (less flash and RAM) is required than for a single-chip solution as there are no applications running on the chip. In the communications industry as a whole, the term network processor refers to any specialized CPU for processing network traffic (e.g. in an IP router). In the low-power wireless industry the term is used to describe SoC network processor solutions with an integrated radio.

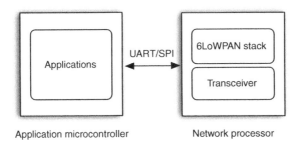

Figure 6.4 Network processor solution architecture.

Communication with a network processor is typically over a UART or SPI interface. This is often realized as an extended socket-like protocol. Thus with this model the use of a 6LoWPAN network requires no integration with the application microcontroller other than the use of a protocol over the local interface. The network processor model is often used also with edge routers, and is easy to integrate with operating systems such as Linux. Edge router integration is discussed in Section 6.4.

The downside of this approach is that two chips are required, which may not be possible for devices with extreme cost limitations. Furthermore, network processors cost slightly more than transceivers as they include a microcontroller, flash and memory.

6.2 Protocol Stacks

The easiest way to enable a wireless embedded device with 6LoWPAN is by integrating an existing protocol stack, either with a network processor, a stack included with an operating system or by integrating a stack into an embedded software project. In this section we look at common open source and commercial protocol stacks which cover all the chip models discussed in the previous section.

A protocol stack for 6LoWPAN typically includes, at the minimum, these basic components:

- radio drivers

- medium access control (e.g. IEEE 802.15.4)

- IPv6 [RFC2460] with 6LoWPAN [ID-6lowpan-hc, RFC4944]

- UDP [RFC0768]

- ICMPv6 [RFC4443]

- Neighbor Discovery [ID-6lowpan-nd]

- socket-like or other API to the stack

Optionally, a protocol stack may include one or more routing protocols, TCP, or various built-in application protocols. 6LoWPAN stacks tend to offer a socket-like API either through library function calls or over a bus interface to a network processor. Application programming for 6LoWPAN will be covered in Section 6.3. Embedded 6LoWPAN stacks are very compact, often taking only 15–20 kB in flash. A white paper discussing some 6LoWPAN stacks and their size is available from the IPSO Alliance [IPSO-Stacks]. The rest of this section gives an overview of two open-source protocol stacks for embedded operating systems: uIPv6 for Contiki and BLIP for TinyOS, and three commercial protocol stacks: Sensinode's NanoStack, Jennic's 6LoWPAN and the Nivis ISA100 stack.

6.2.1 Contiki and uIPv6

Contiki is a popular embedded open-source operating system for small microcontroller architectures such as AVR, 8051 and MSP430, led by the Swedish Institute for Computer Science (SICS) [Contiki]. Contiki includes a very small implementation of IP called uIP [Dunkels03], along with an implementation of IPv6 with 6LoWPAN support called uIPv6. The Contiki operating system and uIP stack are used worldwide by hundreds of projects and companies. The Contiki architecture is designed for supporting IP networking over low-power radios and other network interfaces. The operating system is implemented in C and uses a *make* build environment for cross-compilation on most platforms. A wide variety of microcontroller and device platform ports exist, along with examples and reusable applications.

The architecture of Contiki and uIPv6 are shown in Figure 6.5. The low-level hardware abstraction is split into platform and CPU for portability, which include hardware drivers. The Contiki OS provides basic thread and timer support. The Rime system is a flexible medium access control and network protocol library which includes many low-level communication paradigms. The uIPv6 stack makes use of Rime, and provides a socket-like API for use by applications called *protosockets*. Both built-in and user applications are run over Contiki using a lightweight thread model called *protothreads*.

Contiki is used for the course exercises accompanying this book. More information on developing applications for Contiki can be found in Section 6.3, from the book's course material and from the Contiki documentation [Contiki].

6.2.2 TinyOS and BLIP

TinyOS is an open-source operating system developed for use in wireless embedded sensor network research [TinyOS]. It has widespread support in academia, and includes a large number of experimental communications protocol and algorithm implementations. TinyOS is designed for low-power embedded devices with limited amounts of flash and RAM. The operating system uses a component-based structure and an event-driven execution model. Some object-oriented features are realized by utilizing a new language built upon C which is called NesC. This somewhat limits the portability of the operating system.

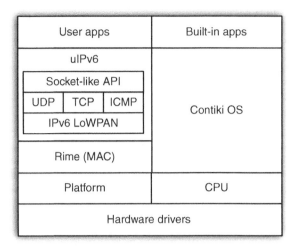

Figure 6.5 The Contiki architecture.

A 6LoWPAN implementation is available for TinyOS from the University of California Berkeley called BLIP (Berkeley IP Implementation) [BLIP]. The blip project is an IPv6 protocol stack including 6LoWPAN header compression, Neighbor Discovery, routing and network programming support. Both UDP and TCP have been implemented along with several application protocols. The project includes Linux support in order to connect TinyOS 6LoWPAN Nodes to other IP networks.

6.2.3   Sensinode NanoStack

NanoStack 2.0 is a next-generation commercial 6LoWPAN protocol stack from Sensinode. This is a compact, optimized solution specifically for 6LoWPAN networking on minimal system-on-a-chip radios. The stack is available as a network processor for SoC radios such as the IEEE 802.15.4 2.4 GHz TI CC2430, CC2530 and the sub-GHz CC1110. Protocol support includes medium access control for the radio chip such as IEEE 802.15.4, IPv6 with 6LoWPAN support, 6LoWPAN-ND, UDP, ICMPv6 and NanoMesh routing. Sensinode's NanoMesh IP routing is based on IETF routing standards and is aimed at large enterprise applications of 6LoWPAN such as smart metering and building automation. The NanoStack architecture is shown in Figure 6.6.

The interface to NanoStack is realized using a socket-like protocol over UART or SPI to the network processor. Thus, unlike typical socket application programming which uses local function calls, the network processor interface is accessed over a local interface. This two-chip solution allows the embedded device microcontroller to stay independent from 6LoWPAN networking and radio communication integration, so that it only needs to implement an interface to the network processor. NanoStack is also available for integration in devices using single-chip or two-chip solutions.

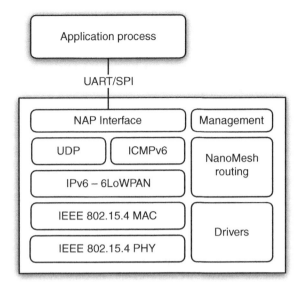

Figure 6.6 The NanoStack architecture.

6.2.4 Jennic 6LoWPAN

Jennic is a wireless chip maker specializing in IEEE 802.15.4 system-on-a-chip radio technology. They were the first chip maker to offer a 6LoWPAN protocol stack for their IEEE 802.15.4 products in addition to their proprietary JenNet and ZigBee protocol stacks [Jennic]. The Jennic 6LoWPAN solution supports 6LoWPAN networking, UDP and ICMPv6 together with an IEEE 802.15.4 MAC. Routing is performed using Jennic's tree-based JenNet Mesh-Under an algorithm which also handles network commissioning and maintenance.

Application programming access is provided through the Jenie API, which is a high-level abstraction for accessing any of the protocol stacks that Jennic offers. In addition, the proprietary simple network access protocol (SNAP), similar to SNMP, is provided for configuring and managing nodes.

6.2.5 Nivis ISA100

The Nivis ISA (NISA) solution is an ISA100.11a protocol stack and system product based on the ISA100.11a standard [Nivis]. The NISA system includes the following components:

- field devices
- backbone router
- gateway
- system manager
- security manager

The system supports two different kinds of network modes: with and without a backbone infrastructure. The radio used in the system is IEEE 802.15.4 compliant, and the NISA100.11a stack is available both on Freescale and Texas Instruments platforms.

6.3 Application Development

This section looks at implementation issues to be considered when developing wireless embedded node applications over 6LoWPAN. The use of Contiki protosockets for application development is introduced. Unlike Internet applications developed for PCs, which are typically for the purpose of personal communications such as web-browsers or VoIP applications, embedded Internet applications are almost always specialized. Embedded wireless nodes usually have a special purpose, with firmware that handles sensing, control, local communication, power savings and other low-level functions in addition to possible user input or output. In order to utilize 6LoWPAN networking, an embedded application uses a protocol stack to configure the network and to send and receive packets. The following non-exhaustive list of practical issues should be taken into account when designing an embedded application for 6LoWPAN networking. In Chapter 5 we discussed general application protocol issues.

Commissioning: In order for commissioning of the link-layer used by 6LoWPAN to be successful, it may be necessary for the application to configure the radio with basic parameters such as the radio channel, data rate, MAC mode and security key to enable basic connectivity with other nodes. The application may also be required to configure the node's EUI-64 or 16-bit short MAC address before using the radio.

Device role: The node needs to configure the protocol stack for the proper type of network role, which is either a host, router or edge router.

Addressing: When using the socket API of a protocol stack, the application may make use of full IPv6 addresses or just MAC addresses depending on the design. Furthermore, if the use of compressed UDP ports is desired, this requires restriction of the port space used by the application.

Mobility: The application may need to deal with some aspects of mobility, for example when the node moves from one LoWPAN to another, causing the IPv6 address of the node to change.

Data reliability: When using UDP as a transport, it provides no guarantee of packet delivery order or reliability. The application must use or implement a protocol that provides sufficient reliability for its purposes.

Security: Security in 6LoWPAN networks is typically supported by the link layer as in IEEE 802.15.4, which provides hop-by-hop encryption. However, at each router the content of the packet is vulnerable as it is decrypted at each hop. Applications that require tight security should implement end-to-end encryption of application data.

As discussed in Section 6.1, there are three different chip solution models. When developing an application using a single-chip or two-chip solution, the application code and protocol stack are on the same microcontroller. In this case, the protocol stack provides a software development API to access the stack. For example, Contiki uIPv6 provides protosockets, TinyOS BLIP provides a component-oriented interface, and Jennic 6LoWPAN provides a high-level interface called Jenie. The integration of an application with a protocol stack on the same microcontroller must be done with care as there are timing, resource and stability issues to be dealt with in addition to portability and evolvability of the application code.

When developing an application that makes use of a network processor chip model, the application runs in a separate microcontroller from the network processor. The application makes use of a local interface to use the networking stack. Sensinode's NanoStack for example provides a socket-like protocol over a UART or SPI interface. The advantage of this approach is that the same interface can be used regardless of the microcontroller architecture, timing and stability issues between the application and protocol stack are decoupled, and the same API can be provided regardless of the radio SoC.

The socket API is the most common programming construct for using IP-based protocol stacks. Figure 6.7 shows an example of the use of sockets from applications (using pseudo function calls). Each application opens a socket construct which is then used to receive or send packets. Each socket is associated with a protocol type (datagram for UDP) and source and/or destination ports.

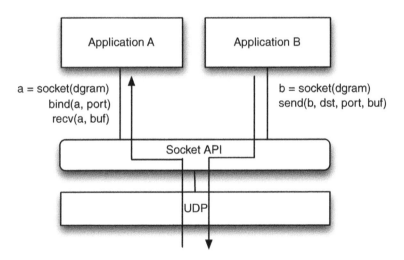

Figure 6.7 Example use of a socket-like API.

The uIP protocol stack for Contiki provides two different interfaces for developing applications over TCP and UDP transports. An event-driven uIP interface is provided, which calls an application function to handle events and retransmissions (optimizing buffer space). Additionally a socket-like API called *protosockets* is provided. The protosocket implementation makes use of Contiki *protothreads*, which are very lightweight threads sharing a common stack. Protothreads enable sequential code resembling standard C-based socket programming. Protosockets have the following basic structure, and must include

PSOCK_BEGIN() and PSOCK_END() calls in order to start and end the underlying protothread. In practice, the socket and its initialization are handled by an application process, and the protosocket thread handles e.g. incoming connections. The following example illustrates how uIP protosockets are used to implement a simple TCP server.

```
static struct psock ps;
static char buffer[10];

/* Declare the protosocket which is called after the socket is
 * created and connected, and a connection comes in.
 */
static
PT_THREAD(handle_connection(struct psock *p))
{
  PSOCK_BEGIN(p);

  /* Send a string over the TCP connection */
  PSOCK_SEND_STR(p, "Welcome!\n");

  /* Close the socket */
  PSOCK_CLOSE(p);

  PSOCK_END(p);
}

PROCESS(example_psock_server_process, "Example protosocket server process");

PROCESS_THREAD(example_psock_server_process, ev, data)
{
  PROCESS_BEGIN();

  /* Listen to TCP port 1010 */
  tcp_listen(HTONS(1010));

  /* Wait for new connections */
  while(1) {

    /* Wait until TCP event comes */
    PROCESS_WAIT_EVENT_UNTIL(ev == tcpip_event);

    /*
     * If a peer connected with us, we'll initialize the protosocket
     * with PSOCK_INIT().
     */
    if(uip_connected()) {

      /*
       * The PSOCK_INIT() function initializes the protosocket and
       * binds the input buffer to the protosocket.
       */
      PSOCK_INIT(&ps, buffer, sizeof(buffer));

      /*
       * We loop until the connection is aborted, closed, or times out.
       */
      while(!(uip_aborted() || uip_closed() || uip_timedout())) {

        /*
```

```
             * We wait until we get a TCP event. Remember that we
             * always need to wait for events inside a process, to let
             * other processes run while we are waiting.
             */
            PROCESS_WAIT_EVENT_UNTIL(ev == tcpip_event);

            /*
             * Here is where the real work is taking place: we call the
             * handle_connection() protothread that we defined above. This
             * protothread uses the protosocket to receive the data that
             * we want it to.
             */
            handle_connection(&ps);
        }
    }
}

  PROCESS_END();
}
```

6.4   Edge Router Integration

In order to connect 6LoWPAN networks to other IP networks, edge router nodes are used.
An edge router typically requires:

- a 6LoWPAN wireless interface,

- 6LoWPAN adaptation,

- 6LoWPAN Neighbor Discovery, and

- a full IPv6 or IPv4/IPv6 protocol stack.

In this section the issues involved with the integration of 6LoWPAN networking in Unix-based edge routers are introduced. In minimal wireless embedded devices, as discussed in the previous section, a 6LoWPAN protocol stack is usually accessed as a library or network processor by network applications on the node. Routing within the LoWPAN is performed in this protocol stack.

The integration of 6LoWPAN networking in edge routers differs greatly in architecture from wireless embedded devices. First of all, because the edge router is connected to a full IP network, it already contains a standard IP protocol stack. Secondly, the edge router has several additional functions to deal with. These include the adaptation of headers between 6LoWPAN and full IPv6, and 6LoWPAN-ND edge router functionality. Finally, the edge router handles IP routing between the 6LoWPAN and IP interfaces and typically also offers firewall, access control and management services for the LoWPAN. Edge routers are often realized using embedded Linux or other embedded operating systems with full IP support, although any PC can serve just as well as an edge router. A Unix-based edge router architecture is shown in Figure 6.8.

One way to integrate 6LoWPAN into an edge router is to provide basic layer 1–3 functionality using a 6LoWPAN network processor, which is used as the wireless interface, as shown in the figure. The 6LoWPAN wireless interface could be realized with a system-on-a-chip radio or a radio module integrated into the edge router hardware – or as a wireless

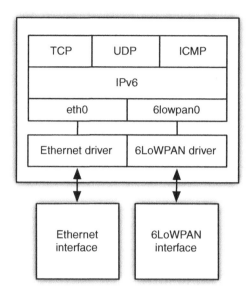

Figure 6.8 Edge router with a 6LoWPAN network interface.

interface peripheral, e.g. in the form of a USB stick. The interface between the wireless interface and the edge router hardware is usually realized with UART, SPI or universal serial bus (USB) interfaces.

To support the wireless interface, a driver is needed for the edge router operating system. This driver implements the interface to the 6LoWPAN stack on the wireless interface. In order to route 6LoWPAN to other IP networks the driver typically emulates a network interface in the operating system, and would show up as e.g. 6lowpan0 in a Unix-based protocol stack. This network interface method is supported for example in Contiki uIP, TinyOS BLIP and NanoStack Linux support. To the standard IPv6 protocol stack this interface looks just like an Ethernet interface, with the exception of a 1280-byte MTU. The IPv6 protocol stack expects to receive standard IPv6 frames from a network interface; thus in order to avoid changing existing IPv6 stacks, 6LoWPAN-related functionality should mostly be implemented below the network interface. Depending on the way that the wireless interface is realized, there may be more or less functionality handled by the OS driver. Some solutions for example implement only IEEE 802.15.4 in the wireless interface, and leave all 6LoWPAN functionality to the driver.

In order to use a 6LoWPAN wireless interface with a standard IPv6 protocol stack, the following functionality needs to be implemented:

LoWPAN Adaptation Layer: The 6LoWPAN frames received from the link-layer need to be decompressed as specified in [RFC4944] and [ID-6lowpan-hc], using known information about the LoWPAN. In the other direction, full IPv6 frames from the network interface need to be compressed. This step could be performed in the wireless interface or the edge router driver.

6LoWPAN-ND: As the edge router acts as a special entity in 6LoWPAN-ND (see Section 3.2), these features are typically implemented below the IPv6 stack although some parts of them could be integrated into or above the stack as well. These features include the 6LoWPAN-ND whiteboard for all 6LoWPAN interfaces and Extended LoWPAN backbone functionality. Additionally, Router Advertisements (RAs) with 6LoWPAN specific options need to be advertised on the 6LoWPAN interfaces. As standard IPv6 stack implementations include RFC4861 ND, the interface or driver should take care of configuring the stack or adapting relevant ND messages between 6LoWPAN-ND and RFC 4861.

With the 6LoWPAN wireless interface functioning as a standard IPv6 network interface, the edge router then implements networking, firewall, access control and management functionality between 6LoWPAN and other IP interfaces. Furthermore, application proxies may also be included on edge routers.

IPv6 routing: IPv6 border routing can be achieved across the 6LoWPAN and IPv6 interfaces by simply using whiteboard information, or by performing route redistribution to another routing algorithm. IP border routing was covered in Section 4.2.7.

IPv6/IPv4 interconnection: LoWPANs can be connected across IPv4 networks utilizing IPv6 transition mechanisms on the edge router or on a router in the local domain. IPv4 interconnectivity was discussed in Section 4.3.

Firewall and access control: As 6LoWPAN devices are very limited, it is usually necessary to provide firewall functionality on edge routers to protect the network from unwanted or malicious traffic. Access control (e.g. in the whiteboard) may be needed to provide network service only to authorized nodes.

Management: Edge routers are ideal places to provide management of a LoWPAN, which can be achieved using standard protocols such as SNMP, or as an HTTP interface.

Proxies: Application protocol proxies, for example to compress/decompress payloads or to adapt protocols, can be implemented in edge routers. The proxy would function on top of the IPv6 protocol stack of the edge router. See Section 5.3.1 for more information on application proxy issues.

7

System Examples

In this chapter we look at how 6LoWPAN is used as a part of complete *systems*. We introduce the ISA100 standard for industrial automation systems, along with two commercial systems that use 6LoWPAN. By analyzing real systems we can put wireless communication and 6LoWPAN networking issues into a practical perspective, looking at where and how 6LoWPAN and related protocols are used. Often the best way to understand the real potential of embedded networking technology is to dig into an application area, which usually reveals a surprising number of uses. The case of facility management introduced in Section 1.1.5 is a good example of this. At the same time, 6LoWPAN networking is only one piece of the embedded system, which often consists of a large number of subsystems and communication technologies. When making a commercial system using wireless IP networking technology, especially when dealing with embedded devices and systems, there are a number of issues that must be faced that aren't necessarily explained in standards – including for example installation cost, ease of use, user privacy and system security.

Industrial automation requires a holistic system approach as there are special requirements for QoS, safety and security. The integration of wireless industrial field devices with back-end supervisory control and data acquisition (SCADA) systems requires careful planning. In Section 7.1 we introduce the ISA100 standard, which defines a complete system solution to wireless industrial automation using IEEE 802.15.4 and 6LoWPAN standards.

Building automation is an important application area for the Wireless Embedded Internet. In the following sections we introduce two examples of commercial systems developed to solve different aspects of automation in buildings. In commercial buildings of all kinds, access control is an important activity for the operator of the building. Traditionally access control was dealt with using people and physical keys, but most modern buildings are equipped with radio frequency identification (RFID) access control systems (also commonly retrofitted in older buildings). In Section 7.2 we introduce a commercial system for RFID-based access control that uses 6LoWPAN to provide a wireless network between access control devices, instead of traditional wiring of all access control components.

In Section 7.3 we look at energy savings in buildings through improved building management and real-time knowledge of how much energy is being consumed by appliances. Governmental studies have shown that a large percentage of energy is consumed as electricity

6LoWPAN: The Wireless Embedded Internet   Zach Shelby and Carsten Bormann
© 2009 John Wiley & Sons, Ltd

or heat in buildings, of which a substantial portion goes to waste. A commercial system using 6LoWPAN networking is introduced that enables real-time information collection about energy consumption and better building management practices.

7.1   ISA100 Industrial Automation

ISA100 is a new standard for wireless industrial automation systems [ISA100]. It is backed by the International Society of Automation (ISA), which is accredited by ANSI. ISA is an international non-profit organization made up of 30,000+ automation professionals. ISA100 is one of a number of standards that include SP95 for enterprise control systems, ISA99 for security, SP50 that documents the Foundation Fieldbus, and a host of other standards for the industrial networking space. ISA100 is not a single standard, but is expected to be part of a family of standards designed to support the wide range of wireless industrial plant needs to include process automation, factory automation and RFID. The core wireless automation system of ISA100 is standardized as ISA100.11a.

Specifically for *non-critical* monitoring, alerting and supervisory control, ISA100.11a has been designed to support these and the needs of process control where latencies on the order of 100 ms can be tolerated. The standard defines the protocol stack, system management and security functions for use over low-power, low-rate wireless networks (currently only the IEEE 802.15.4 standard). 6LoWPAN standards are utilized for networking.

The design goals for ISA100.11a were:

- an open standard

- simple to use and deploy

- serve industrial and process control

- assure multi-vendor interoperability

- support and utilize existing open standards

- coexist with existing installed wireless networks

7.1.1   Motivation for industrial wireless sensor networks

Numerous protocols and networks already exist in the industrial process control space such as Foundation Fieldbus, Profibus, HART, CIP and others. Most all of these use various forms of wired networks to attach the sensor and control/actuator devices together. The cost of installing and maintaining the wiring for these networks is becoming prohibitive, especially as the size of the plants grow and the operators wish to increase the number of monitor points. In one case, the industrial plant that was used as part of the use-case requirements was over 20 square miles in area and had tens of thousands of sensor points. The operators of these plants want a lower-cost (calculated over the total cost of ownership – TCO) solution as compared to wired networks. The TCO in this case includes the devices, physical media, cost of installation, cost of commissioning, device and media maintenance, and ease of expansion. Additionally, they require interoperability between devices from different vendors as well as coexistence with existing networks and technologies. A standards-based wireless solution would provide this lower cost, interoperability and also the ability to greatly increase the number of control and sensor points, providing more information to plant operations which will improve the productivity and safety of the plant.

There are two distinct application areas within industrial automation – process control and factory automation. Process control will typically deal with petroleum, chemical or gas production, and factory automation with manufacturing such things as appliances, discrete components and consumer products. The typical sensor network application for both segments will be for data collection including device monitoring – for example sensing motor vibrations or temperature – and asset tracking. Systems will be deployed to capture some process monitoring for non-critical process control events, such as environmental temperature control and recording for regulatory compliance. It isn't expected that these wireless sensor nodes will replace existing wired nodes or be used in critical process control applications. They will instead be used to augment the condition monitoring and for expansion to include controlling devices that were not previously connected.

To further ease the deployment and increase the number of installed sensors, support for long battery-life is required. While these devices may transmit messages at an interval of one per second, more typically the average will be one message per minute with only tens of bytes of data per message. Additionally, some devices may transmit stored log files or time-series data on a daily basis that could be tens of kBs of data. Multiple wireless networks may be deployed within the same location with overlapping RF communications. In most cases these will be distinct and separate networks with no need to communicate between nodes on each.

7.1.2   Complications of the industrial space

The industrial space provides some unique challenges for wireless sensor networks. Within the industrial market there are six different classes of sensor and control applications ranging from *critical safety* (class 0) to *condition monitoring* and *regulatory compliance* (classes 4 and 5); Table 7.1 gives an overview and some examples of the various types of applications for each class. ISA100 was designed primarily for class 4 and 5 applications, but the architecture will also provide support for class 2 and 3 (control) applications. The major difference between the classes is the latency and timing requirements. As the class numbers decrease the latency, timing and jitter requirements increase, i.e. only small latencies and jitter can be tolerated.

The need for lower and tighter controlled latencies while at the same time providing support for bursty bulk data transfers over shared media necessitates the use of some form of bandwidth allocation over the available channels and time. A typical *closed loop* control system will require latencies around 10 milliseconds with the added constraint that if packets are not received "in time" the process system will initiate a shutdown. In an *open loop* (human intervention) control system the required latency needs to be below 150 milliseconds to provide the necessary responsiveness. It was felt by the ISA100 design team that a simple CSMA/CA algorithm, typically used by 802.15.4 devices would not suffice. A more complex graph routing based on time division multiple access (TDMA) with bandwidth and throughput contracts was used instead. This was also chosen in part because the typical traffic uses a "publish/subscribe" model where data is sent (published) from the sensor devices to one or more systems (subscribers). The application layer publish/subscribe paradigm was discussed in Section 5.3.3.

Table 7.1 Classes of sensor and control applications.

Class 0	Safety	Emergency action required
		• Emergency Shutdown • Automatic Fire Control • Leak detection
Class 1	Control	Closed loop control – Critical
		• Direct control of actuators, pumps and valves • Automated shut-down
Class 2	Control	Closed loop control – Non-critical
		• Optimizing control loops • Flow diversion
Class 3	Control	Open loop control – Human intervention
		• Operator performs manual adjustment
Class 4	Monitoring	Alerting – Necessary maintenance
		• Event based maintenance • Low battery • Vibration monitoring • Motor temperature monitoring
Class 5	Monitoring	Logging – Preventive maintenance
		• Preventive maintenance records • History collection

7.1.3   The ISA100.11a standard

The ISA100.11a committee reviewed available protocols before embarking on the design of a new protocol. Specifically the group looked at Wireless HART and ZigBee as potential solutions, but both had significant shortfalls. Some of the design criteria included:

- reliability (enhanced error detection, frequency hopping)

- determinism (TDMA, QOS support)

- security

- optimization for sensor data flows

- support for multiple protocols

- support for multiple applications

- use of open standards

Whereas wireless HART was also designed for industrial wireless sensor networking applications and does provide some of the same features as ISA100.11a, it does not support multiple protocols. Wireless HART only specifies and supports the transmission of HART messages over a wireless physical media. Wireless HART and HART were designed for process control. They were not designed to support additional applications such as factory automation.

ZigBee supports wireless control and battery-powered operations, but it does not provide the necessary QoS support to enable the latency and message flow determinism required for industrial applications. ZigBee also only supports ZigBee messages and would not be able to support other protocols such as HART, Modbus, Foundation Fieldbus and Profibus. And finally, ZigBee was developed and is maintained by a closed special interest group (SIG) and is not an openly available international standard.

There are numerous other proprietary network architectures and protocols that have been deployed and are being deployed, but they generally fail to meet some of the design criteria and specifically fail to meet the requirement of being an open standard.

A typical ISA100.11a network is shown in Figure 7.1. The network design in the figure attempts to show the tree nature of the ISA100.11a architecture. Since much of the traffic is expected to flow from bottom to top in the figure, a tree provides the most efficient topology for such a network. Note that the ROLL routing protocol covered in Section 4.2.6 assumes a similar topological design. There is little or no requirement for node-to-node communications, except to forward packets to the plant network and the control system. The solid and dashed lines in the figure show multiple routing graphs supported by this network.

ISA100.11a is based on international standards. The PHY and MAC layers are version 2006 of IEEE 802.15.4 and the network and transport layers are based on 6LoW-PAN [RFC4944, ID-6lowpan-hc, ID-6lowpan-nd], IPv6 [RFC2460] and UDP [RFC0768] standards. The ISA100 data link layer implements the graph routing and TDMA features of the standard. The forwarding of messages within the wireless network is performed at the link layer, i.e. using a link-layer Mesh-Under design – see Figure 7.2. Link-layer mesh forwarding under the LoWPAN adaptation layer was discussed in Section 2.5. Future versions of ISA100 may also support the ROLL routing protocol, as it has been designed to support similar requirements in [ID-roll-indus].

Because ISA100.11a leverages IPv6 protocols and addressing, it also uses similar terminology. All of the nodes connected within a single link-local mesh or PAN are collectively called a *DL subnet* (data link subnet). As can be seen in the figure, packets are forwarded between nodes at the ISA100.11a data link layer. Until a packet reaches either the destination node within the DL subnet or the border router, it does not get interpreted by the LoWPAN adaptation and IP layers. Messages are forwarded within the DL subnet transparently to the upper layers. As a result, the ISA100.11a data link layer provides an abstraction of a broadcast-type network to the higher layers. The ISA100.11a network supports:

- mesh, star-mesh and star topologies

- non-routing sensor nodes

- connection to a plant network via a gateway

- device interoperability

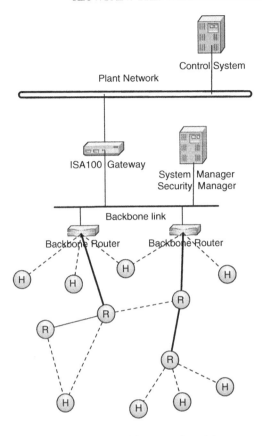

Figure 7.1 The ISA100 network architecture.

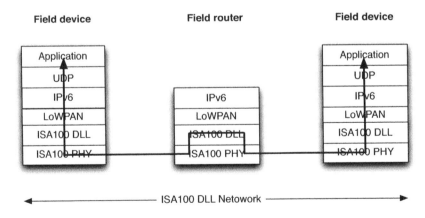

Figure 7.2 Forwarding at the link-layer through the ISA100 protocol stack.

- data integrity, privacy, authenticity, replay and delay protection

- coexistence with other wireless networks

- robustness in the presence of interference

- networks with up to 30,000 nodes

7.1.4 ISA100.11a data link layer

As mentioned, the ISA100.11a DLL provides support for the creation, maintenance and packet forwarding functions required for the wireless sensors. In the OSI model the DLL sits between the physical layer and the network layer. It establishes data packet structure, framing, error detection and bus arbitration. The DLL also includes the medium access control (MAC) functions. In ISA100.11a the DLL was extended to include the following functions:

- link-local addressing for PAN entities

- message forwarding from one PAN node to another

- PHY management

- adaptive channel hopping

- message addressing, timing and integrity checks

- detection and recovery of message loss

Messages are communicated in discrete 10–12 millisecond time-synchronized slots. This time synchronization provides extremely accurate time stamping, and the adaptive channel hopping increases reliability by avoiding occupied or noisy channels. In addition, the time-synchronized slots and channel hopping reduce the utilization of any single channel thereby improving ISA100.11a's coexistence with other RF networks in the same spectrum.

The DLL creates and uses a tree-like routing algorithm called graphs. The concept of graph routing provides for a number of different data paths for different types of network traffic within the DL subnet. The multiple graphs are used by different nodes to transmit different types of data. For example, a node will use one graph route to send intermittent sensor data to the plant network and will use another graph route to send bulk series data. While all of the roots of the graphs are generally anchored at the same point, within the DL subnet, the path of the messages from any single node may vary greatly depending on the traffic type, bandwidth required and other factors. These various graphs are created by the network management system (NMS).

The NMS takes input from the system/network designer as to the specific requirements for data throughput and transmissions, and information provided by the sensor nodes about the RF environment. Based on this the NMS calculates and creates a set of graphs, and assigns contract IDs to the graphs. The applications on the sensor nodes then use these contract IDs to notify the nodes on the route to the plant network as to the requirements for the transmission and forwarding of that particular message. The graphs are instantiated based on specific traffic characteristics. The ISA100.11a committee defined four traffic types:

Periodic data: Data that is published periodically and has a well-understood data bandwidth requirement, both deterministic and predictable. Timely delivery of such data is often the core function of a wireless sensor network, and permanent resources are assigned to ensure that the required bandwidth stays available. Buffered data usually exhibits a short Time to Live, and the newer reading obsoletes the previous one. In some cases, alarms are low priority information that get repeated over and over. The end-to-end latency of this data is not as important as the regularity with which the data is presented to the plant application.

Event data: This category includes alarms and aperiodic data reports with bursty data bandwidth requirements. In certain cases, alarms are critical and require a priority service from the network.

Client/server: Many industrial applications are based on a client/server model and implement a command-response protocol. The data bandwidth required is often bursty. The acceptable round-trip latency for some legacy systems was based on the time to send tens of bytes over a 1200 bit/s link (hundreds of milliseconds is typical). This type of request is statistically multiplexed over the network and best-effort service is usually expected.

Bulk transfer: Bulk transfers involve the transmission of blocks of data in multiple packets where transient resources are assigned for a limited period of time (related to file size and data rate) to meet the bulk transfer's service requirements, such as a transaction time constraint.

The NMS uses these different traffic types along with the requirements for the amount of data, frequency and latency requirements to calculate primary and backup graphs for the specific traffic. The NMS also takes into account the performance of the various channels between pairs of nodes as well as the power constraints of each node that the traffic will pass through. At the end of the calculations contract IDs are assigned to each of the graphs. These contract IDs are carried within the DLL header, and each node on the forwarding tree examines the contract ID to determine the "next hop" for the message.

In this way the ISA100.11a DLL provides a robust and flexible network topology upon which to build higher layer network functionality. It supports low-power and high-availability operations even in the presence of non-intentional interferers and it supports data transmission characteristics for the different types of network traffic required in an industrial process control or factory automation application.

7.2   Wireless RFID Infrastructure

A system for deploying RFID reader infrastructures for identification and access control using wireless 6LoWPAN technology has been developed by Idesco and Sensinode. Idesco is a Finnish manufacturer of identification systems, and was the first company in the world specialized in RFID [Idesco]. Idesco offers scalable open identification solutions enabling new business opportunities for forerunners in the field. This new wireless system is called the Idesco Cardea system, consisting of a line of wireless RFID products including:

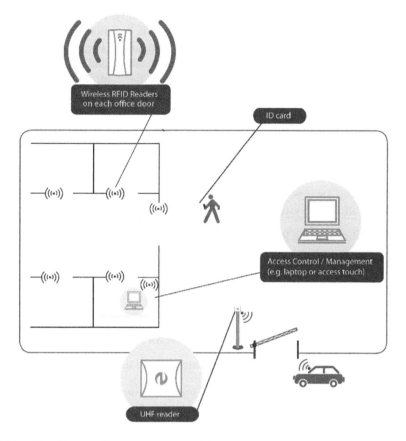

Figure 7.3   The Idesco Cardea system architecture. (Reproduced by Permission of ©
Idesco Oy.)

- Idesco Cardea readers

- Idesco Cardea door control unit

- Idesco Cardea control unit and access touch

The system uses Sensinode wireless networking technology, based on 2.4 GHz IEEE
802.15.4 and 6LoWPAN standards. Traditional RFID tags are read at short range using
the RFID Cardea reader products. Instead of using wired cabling between all of the RFID
infrastructure components, the Cardea system uses a wireless 6LoWPAN mesh network to
interconnect the devices. The system is illustrated in Figure 7.3. A typical system consists of
RFID readers, door control units and a controller unit. The Cardea system includes both a
Cardea controller unit accessible through PC software, or the Cardea access touch, which is
a wall-integrated touch-screen device.

Traditional RFID infrastructures rely heavily on cabling between all components in the
system and the use of legacy centralized controller units. Cabling is typically realized in these

systems using RS-485, Wiegand connections or, more recently, IP-based Ethernet cabling. The Wiegand protocol (named after the Wiegand effect) is commonly used for connecting card readers with electronic entry systems over three-wire cables. The installation of cabling, especially in existing and historical buildings, is a painstaking and expensive investment. The use of legacy cabling has also resulted in the centralized control of these systems with specialized legacy controllers. The combination of cabling and legacy infrastructure makes RFID systems expensive, and limits the use of RFID in many applications. By using a low-power wireless mesh network, several benefits were achieved with the Idesco Cardea system:

- The Idesco Cardea system takes just an estimated 40 minutes to install per door, compared to an average of 7 hours per door for a traditional wired RFID infrastructure.

- Significant cost savings in cabling for retrofitting the system in existing buildings, or for modification of a system in a current installation.

- The possibility of using RFID solutions as part of temporary installations.

- RFID becomes practical and cost-effective for small office installations.

- The ability to easily combine existing wired and Cardea wireless components.

Typical applications of the Cardea systems include:

- small installations with a few doors and a few readers (such as small offices or shops)

- temporary access systems, e.g. at construction sites or events

- systems where multiple applications need to be combined together (e.g. access control and parking lot access)

- applications with readers spread across large areas, such as harbors

- installations in existing and especially historic buildings

- easy expansion of existing systems without adding more cabling

7.2.1   Technical overview

The system is realized using Sensinode's NanoStack 2.0 6LoWPAN solution, providing communication between the wireless components in the system as shown in Figure 7.4. Making use of IEEE 802.15.4 2.4 GHz radios with power amplification, suitable range is achieved. Furthermore, multihop routing using Sensinode's NanoMesh is used to extend the network over larger areas. Security is achieved with 128-bit AES link-layer encryption combined with application protocol security already used in wired systems. Devices in the network use unique 64-bit MAC addresses so that all devices have a unique IPv6 address. In the future, battery-powered wireless RFID readers will be added to the cardea system. These run as LoWPAN hosts with aggressive power savings for long battery life. All powered devices, including RFID readers and door controller units act as LoWPAN Routers. The controller units in the network act as LoWPAN Edge Routers. The network achieves complete autoconfiguration, which makes the setup of the system using a graphical interface easy. Installation of the system doesn't require special knowledge of IPv6 networking or wireless

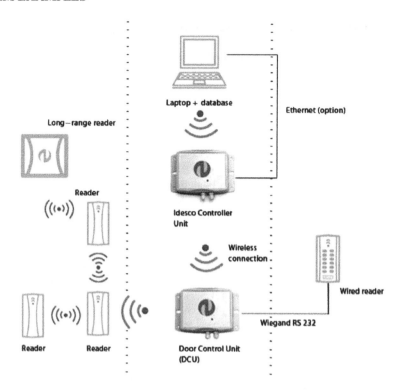

Figure 7.4   The wireless communications between Cardea components. (Reproduced by Permission of © Idesco Oy.)

communications, and can be performed by existing installation personnel. The use of IPv6 end-to-end will enable the evolution of the Cardea system into larger-scale access control applications and integration with other building automation systems.

7.2.2   Benefits from 6LoWPAN

6LoWPAN technology has enabled Idesco to continue its leading developments in the RFID field. The resulting Idesco Cardea system is unique in the world. "Idesco sees a great market potential for wireless Cardea products. 6LoWPAN technology, which is designed for transferring small data amounts over short distances, is ideal for applications using RFID for identification. The unique identification number (UID) is typically the only data that needs to be transferred from the RFID reader to the upper levels in the system. As the network routes itself automatically and no cables are required, the installations can be done in a more flexible and cost-efficient manner." says Anu-Leena Arola (Director, R&D and Services) for Idesco.

7.3   Building Energy Savings and Management

The LessTricity system has been developed by a consortium of companies in the UK, for the purpose of increasing the efficiency of energy usage and better management in commercial buildings and businesses. The consortium includes companies involved in commercial property management (MEPC), building design, engineering and management (WSP Group), electronic product design and manufacturing (GSPK Design Ltd), management software design and implementation (TWI) and low-power wireless networking (Jennic Ltd). The project has received additional funding from the UK government's Technology Strategy Board whose mission is to promote and support research, development and exploitation of technology and innovation for the benefit of businesses in the UK.

The aim of the LessTricity consortium was to develop a centralized management system to help eliminate the wasteful use of electrical power by appliances in buildings. The system uses wireless control and measurement technology based on 6LoWPAN to enable the management of large facilities and even remote buildings. It can be easily installed in both new and existing buildings, is transparent to users, offers advanced metering features and has low operating costs. Such a system can also be integrated into a larger facility management system, such as that introduced in Section 1.1.5.

LessTricity technology is initially aimed at industrial, commercial and public service market segments, which make up almost 52 percent of electricity consumption in the UK. A report in 2004 from the UK Department of Environment, Food and Rural Affairs showed that UK businesses wasted 30 percent of energy purchased and, for many, a 20 percent cut in energy costs would represent a benefit equivalent to a 5 percent increase in revenue given rising energy costs [DEFRA].

7.3.1   Network architecture

The network architecture of the LessTricity system is shown in Figure 7.5. A typical small deployment consists of clusters of about 50 LessTricity power controllers (LPCs) into which appliances to be measured are plugged. Each LPC monitors the amount of power being consumed and transmits this through a multihop 6LoWPAN network to a LessTricity network interface (LNI). The LNI acts as a LoWPAN Edge Router between the LoWPAN and the Ethernet network. Power readings from LPCs are sent to a central database where they are stored. Standardized reports on aggregate and individual appliance energy use can be produced using SQL queries, to allow building owners to analyze their energy consumption. A graphical web interface is also provided for both monitoring and managing energy consumption in the LPCs and to manage the network and LPCs themselves.

7.3.2   Technical overview

The LPCs consist of an enclosure containing a standard UK 13 A socket into which the appliance is plugged, and a lead terminating in a 13 A 3-pin plug for connection into the mains power supply. The device is able to measure instantaneous current and voltage on the socket, attached to a wireless microcontroller module supplied by Jennic.

The LPCs communicate back to the wireless infrastructure of the building using the transmitter of the JN5139 system-on-a-chip radio. This single-chip device contains a 32-bit processor along with peripherals and an IEEE 802.15.4 2.4 GHz radio. Thus this device

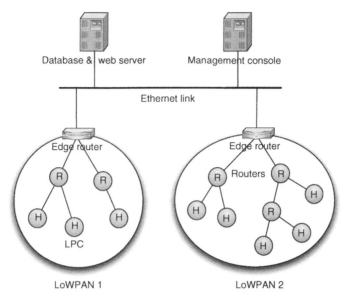

Figure 7.5 The typical network architecture of a LessTricity deployment.

has been implemented using the single-chip stack model described in Section 6.1.1. The protocol stack running on the LPCs and LNI is illustrated in Figure 7.6. Multihop networking is provided by Jennic's JenNet self-healing Mesh-Under technique over the IEEE 802.15.4 MAC. Thus this can be considered to be a proprietary link-layer mesh technique. A LoWPAN adaptation layer compliant with [RFC4944] is implemented under IPv6, along with UDP and ICMPv6 implementations. Any LPC is capable of acting as a router for packets from any other device. Jennic's simple network access protocol (SNAP), covered in Section 6.2.4, is used as an application protocol for the LessTricity application running on the nodes.

The LNI is based around Jennic's Ethernet border router consisting of a higher performance microcontroller with an Ethernet interface along with a JN5139 wireless module, which are interconnected by a high-speed serial interface. This configuration follows the network processor setup for edge routers described in Section 6.4. An IPv6 protocol stack on the Ethernet microcontroller handles networking on the Ethernet side and routes packets between the two interfaces. The JN5139 handles 6LoWPAN networking and acts like a network interface under the IPv6 stack.

7.3.3 Benefits from 6LoWPAN

By using a 6LoWPAN based approach, the system could be easily deployed in a building using the low-power IEEE 802.15.4 wireless network, while at the same time connect into the building's existing IT infrastructure thanks to the use of standard Internet protocols. The energy usage information can then simply be made available locally or over the Internet for remote monitoring, e.g. for corporate energy usage monitoring. The end-user focus is on reducing the energy footprint of their operations, both from a cost savings and an

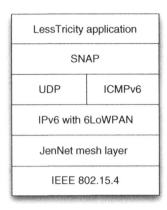

Figure 7.6 The Jennic 6LoWPAN stack with the LessTricity application.

environmental responsibility point of view. "We have found it a great advantage to be able to add Ethernet connectivity to a wireless network and with it the possibility of Internet access to any node simply by utilizing existing IP-based standards. Coupled with the SNAP layer, it has enabled several different applications to be written and deployed for Internet use in very short timescales," says Paul Chilton of Jennic.

The LessTricity system has gone through field trials in 2009 in business facilities around the UK. These field trials were used to establish a base energy usage statistics for the trial sites, which can be compared to their energy usage after the management features of the system are taken into use.

8

Conclusion

The next billion nodes on the Internet will not be PCs, laptops or "netbooks": they will be embedded systems solving a large number of problems in industry, building management, utilities, energy management and home comfort. Many of these will use wireless connectivity, and many will be low-power.

The success stories of the Internet, or of technologies such as Ethernet or 802.11 WLAN (WiFi) show that one of the most important success factors for a networking technology is ease of integration. What needs to be done to make wireless embedded networks as easy to integrate as today's Ethernet or WLAN products?

The 6LoWPAN concept of an edge router makes sure LoWPANs integrate easily with the majority of wired and wireless networks that will be used in future deployments. There is no need to change the edge router for each new application; the network is concerned with *networking*. So the integration problem can be considered solved at a packet transport level. However, this is not the whole story:

- The processes for the setup and management of the networks have to work together. In particular, the security models have to fit, and the setup of a LoWPAN must not require the setup of a whole new world of security parameters and attributes. In general, it is a challenge to reconcile the requirements of security and usability.

- LoWPAN Nodes (and edge routers) of different vendors have to integrate easily. The integration may not be completely "plug and play", as most systems need some commissioning (as discussed in Section 3.1), but it should be painless. This requires further standardization in the areas of security, routing and management.

- Applications need to integrate with nodes of different vendors. While it is unlikely (and probably actually not that desirable) that the whole world converges on a single application protocol, it is very much desirable that there is a small set of "default choices" for application layer protocols for each specific area of application.

Vertical approaches such as the ZigBee branded series of specifications provided full-stack solutions that solved (or at least could have) all the internal integration problems between

6LoWPAN: The Wireless Embedded Internet   Zach Shelby and Carsten Bormann
© 2009 John Wiley & Sons, Ltd

the elements of the technology, at the expense of leaving the integration between the new island and the existing IP world unsolved. The standards-based approach of the Wireless Embedded Internet has already solved many of the latter issues, but needs work on full-stack and system solutions. Standards such as ISA100 show an interesting way forward: to build system standards out of standards-based components, enabling easy integration and possibly plug-and-play operation by supplying "glue" for a specific area of application. The ETSI M2M standardization effort is another good example, where leading industry players involved with machine-to-machine systems (including smart metering) are standardizing the entire system from embedded nodes and gateways, to access networks, back-end systems and the interface to applications.

Clearly the IETF is not in the systems standards business. The IETF's objective is *to make the network work*. The IETF will always seek alliances with more vertically oriented organizations such as ISA to complete the stack. However, IETF standardization can pave the way for more interoperability in at least three areas:

Routing: It is unlikely that a single routing protocol can be a perfect fit for the wide variety of applications envisioned for the Wireless Embedded Internet. The Internet operates with a variety of routing protocols within its specific domains. To achieve plug-and-play operation with this divergence, the host-router interface is used, which has been designed to be independent of routing protocols.

However, LoWPAN Routers are commodity parts almost as much as LoWPAN hosts. There should be an expectation of plug-and-play for the LoWPAN Routers as well. But one size may not fit all.

The IETF ROLL WG has collected detailed requirements for four fields of applications and is now in the process of designing a solution (see Section 4.2). This solution will work best if it is modular, with a modest base protocol and options that incrementally increase the level of service if deployed on some (but not all) of the routers, creating true interoperability.

Security: The IETF has an impressive track record of creating protocol standards with real security, not least because the early Internet did so little in the way of security, and the transition has been painful. Other standards organizations (such as the IEEE and lately the XMPP Foundation) have gone to the IETF to hammer out robust security protocols.

The IETF approach to security is to supply building blocks that can be combined into the system needed for a specific application. Many of the existing building blocks will be very useful for the management/commissioning side of the security process; Section 3.3 has shown that other building blocks such as IPsec's ESP may be directly applicable as well.

One of the challenges will be coming up with mechanisms that bind as little policy as possible into the little nodes in the LoWPAN while still making it possible to implement a large variety of policies at appropriate control points elsewhere in the system.

Another challenge will be to resist the temptation to limit the architectures to single control points. Of course, any organization would like to "own" the network, the applications and the customer. A painful experience of the last decade for many organizations used to this level of control was that this no longer works. No single vendor will control an entire plant network. Networks need to be open and provide for choice. Designing open security mechanisms is a challenge, but has been done successfully.

Application protocols: Although 6LoWPAN enables networking over low-power wireless embedded networks, it does not enable successful use of the technology unaided. This network-related work must be complemented by vertical standardization. Suitable application protocols and data formats for use over 6LoWPAN networks, and end-to-end over the IPv6 Internet, are going to be an important area of development in the future. While the IETF *could* define a protocol and data format for reading temperature sensors, this is best left to organizations aligned with the specific needs of a segment of application.

However, end-to-end application protocols and formats for different areas of application could still share some basic properties that could be standardized by an organization like the IETF – as an analogy, the IETF does not standardize "the Web", but it does work on HTTP and cooperates with the W3C on important formats. It would certainly help with market acceptance if 6LoWPAN had a "killer application protocol", like HTTP has been for the most popular use of the Internet, the World Wide Web. But as embedded applications are so heterogeneous, there will no doubt need to be a range of protocols – no one-size fits all.

In Chapter 5 we identified the state-of-the-art in application protocols, along with suitable candidates for use with 6LoWPAN. It can be concluded that although there are several protocols which can already be used with 6LoWPAN, this is just a start. Many approaches need optimization, compression and further research. ZigBee's application profiles provide a good starting point, as they already solve part of the problem for ad hoc networking in specific domains. Web services are also seen to be such a widespread and powerful technique, that embedded web services suitable for 6LoWPAN will surely be a success. We see a need for work in this area, and hope to see activities in the standardization of application protocols and data formats suitable for 6LoWPAN in the future both at the IETF and elsewhere. Recently a new standardization effort with these aims has been started at the IETF called 6lowapp (see http://6lowapp.net).

One question related to this is how the existing full-stack islands will continue to evolve. Recently, the ZigBee Alliance made waves with an announcement entitled: *ZigBee Alliance plans further integration of Internet Protocol standards* [ZigBee]. Unlike the previous approach of ZigBee, which has used an application gateway to bridge between proprietary ZigBee networks and IP networks, the new announced approach appears to be very different. The alliance plans on integrating IP standards directly into the network stack of future specifications of the ZigBee series, thus greatly increasing cooperation with the IETF and the IPSO Alliance and accelerating the convergence to IP happening across the board in the embedded networking industry. Although we don't know exactly how IP standards will be integrated with ZigBee, the work on the ZigBee compact application protocol (CAP, covered in Section 5.4.3) gives us one option. The logical way to integrate IP into one of the ZigBee stacks is to replace the ZigBee network layer with 6LoWPAN, IPv6 and ROLL routing and the ZigBee application protocol with UDP, as specified in CAP. However, the upcoming ZigBee/IP Smart Energy specification will make use of open IETF, W3C and IEC standards rather than the ZigBee application layer over CAP. The ZigBee brand is relatively well known, to the point that some literature confuses ZigBee with the entire IEEE 802.15.4 genre. It remains to be seen how well the transition from a full-stack island to an open standard will work for the ZigBee brand and organization.

The Z-Wave networking technology developed by the vendor Zensys and supported by an industry SIG makes use of proprietary sub-GHz radio technology and its own specialized networking. So far, the connection of Z-Wave islands, usually utilized in home automation, with the Internet has been solved with gateways, using a similar approach to ZigBee. With 6LoWPAN standards now open to a wider variety of link-layer technologies, Z-Wave will be a suitable candidate. The benefits of such an integration are immense, but at the same time a challenging transition for an island technology with a large number of existing legacy users.

Finally, it helps to have organizations devoted to making things work also on the commercial marketing, public relations and interoperability of IP technology. Right now, the IPSO Alliance is most active in the commercial promotion of IP technology in embedded systems. Already, early in its existence, it has managed to round up an impressive range of promoters and members ranging from low-power RF chip makers like Texas Instruments and Atmel, the networking giant Cisco, computing companies like Sun Microsystems and the ubiquitous Google. We see IPSO playing an important role in making the Wireless Embedded Internet a success, as it fills many gaps in which the IETF is not active.

Although enough work remains to be done, 6LoWPAN is in an excellent position to be an important building block for the Wireless Embedded Internet. There are challenges in the technology, but there are also challenges in getting the different cultures of the radio, networking and various application domains together. We don't want to repeat the 1990s *bellheads vs. netheads* debate; we want to *make the network work* and deploy billions of wireless nodes that solve real problems.

A

IPv6 Reference

IPv6 is the successor to the wildly successful IPv4 Internet Protocol. This section provides a quick reference to some header formats and other important features of IPv6 that are useful as reference material while perusing the main body of this book.

A.1 Notation

Figure A.1 shows an IPv6 header in graphical form. The form has been chosen to emphasize the new focus of IPv6 on 64-bit wide data structures.

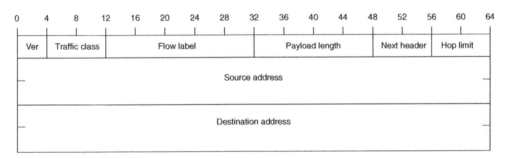

Figure A.1 IPv6 packet header.

The most relevant documentation for Internet standards is in the form of RFCs, a 40-year-old series of documents traditionally exchanged as ASCII line printer format pages. The column width of these documents is limited to 72 characters, making it hard to represent 64-bit wide data structures in *ASCII art*. The traditional way to represent packet structures in RFCs therefore limits itself to rows of 32 bits, yielding ASCII art that looks like Figure A.2.

Each tick mark represents one bit position. The numbers at the top count the bits from bit 0, the most significant bit in the first byte (generally the most significant byte as most Internet protocols use network byte order) to bit 31, the least significant bit in the fourth byte. (Note

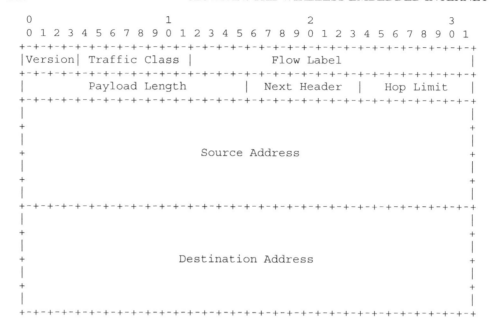

Figure A.2 IPv6 packet header in box notation.

that the 0/1/2/3 above the numbers line are the digits for the tens, they *don't* indicate byte
boundaries!)

This notation, sometimes affectionately called *box notation*, is the notation that can be
found throughout the most important RFCs documenting the standards that govern the
Internet. Instead of trying to come up with our own notation for this book, we decided to
simply stick to the standard notation, in the hope that this will make the figures much easier
to recognize when consulting the actual reference documents.

As a slight simplification, we use underscores to identify reserved fields of packet
structures, as can be seen in Figure B.1 in the next appendix. Reserved fields usually have to
be sent as zero bits; the rules on what to do when received as non-zero differ from structure
to structure – the reader is advised to consult the RFCs for the fine details.

A.2   Addressing

The main new feature of IPv6 compared to IPv4 is the new address size, based on 128-bit
addresses. It would be unwieldy to continue to write them in byte-wise decimal notation as
we do for IPv4 addresses. Instead, the notation for IPv6 addresses is based on eight 16-bit
values separated by colons, each 16-bit value represented by up to four hex digits (leading
zeros are usually omitted). As a further abbreviation, *one* sequence of all-zero 16-bit values
can be replaced by a double colon indicating a longer sequence of zeros, as illustrated in
Table A.1.

Table A.1 IPv6 addresses in hexadecimal notation.

Long form	Abbreviated form	Explanation
2001:DB8:0:0:8:800:200C:417A	2001:DB8::8:800:200C:417A	a unicast address
FF01:0:0:0:0:0:0:101	FF01::101	a multicast address
0:0:0:0:0:0:0:1	::1	the loopback address
0:0:0:0:0:0:0:0	::	the unspecified address

```
| 10 bits  |         54 bits            |           64 bits            |
+----------+---------------------------+------------------------------+
|1111111010|             0             |         interface ID         |
+----------+---------------------------+------------------------------+
```

Figure A.3 IPv6 link-local address.

```
| 3 |      45 bits        |  16 bits  |           64 bits            |
+---+---------------------+-----------+------------------------------+
|001| global rting prefix | subnet ID |         interface ID         |
+---+---------------------+-----------+------------------------------+
```

Figure A.4 IPv6 global unicast address.

The initial bits of an IPv6 address group the address into one of several address formats, which are summarized in this section. The full details of the IPv6 addressing architecture are documented in [RFC4291], and the global unicast address format in [RFC3587].

Link-local IPv6 unicast addresses. Link-local addresses used for bootstrapping and other communication confined to a single link. They start with FE80::/10. The format of link-local addresses is shown in Figure A.3.

Global unicast addresses. The general format for IPv6 Global Unicast addresses is shown in Figure A.4.

The IPv6 address architecture document allows a number of different values for the first three bits of a global unicast address. As of now, only one format has been put into general use: addresses starting with 001 binary, or with the 2000::/3 prefix (i.e. the first hex digit is a 2 or a 3).

These three bits, together with the next 45 bits, form a /48 prefix that is meant to be assigned to a site (a cluster of subnets/links). The subnet ID is an identifier of a link within the site, and the interface ID identifies a specific interface (and thus a node) on this link.

Multicast addresses. An IPv6 multicast address identifies a group of nodes (more precisely: a group of interfaces, usually on different nodes). Each IPv6 interface may belong

```
|   8    | 4  | 4  |                   112 bits                       |
+--------+----+----+--------------------------------------------------+
|11111111|flgs|scop|                   group ID                       |
+--------+----+----+--------------------------------------------------+
```

Figure A.5 IPv6 multicast address.

```
  0 1 2 3
  +-+-+-+-+
  |0|R|P|T|
  +-+-+-+-+
```

Figure A.6 Flag values for IPv6 multicast addresses.

Table A.2 Scope values for IPv6 multicast addresses.

Hex value	Scope
1	Interface-local scope
2	Link-local scope
4	Admin-local scope
5	Site-local scope
8	Organization-local scope
E	Global scope

to any number of multicast groups. The format of multicast addresses is shown in Figure A.5.

The flags (flgs) field provides four bits of options which are assigned as shown in Figure A.6. The T bit indicates that the multicast address is of a temporary nature, the P bit identifies the address as a unicast-prefix-based IPv6 multicast address [RFC3306], and the R bit indicates that the rendezvous point (RP) address is embedded in the IPv6 multicast address [RFC3956].

The scope (scop) field is a four-bit value that specifies the scope of the multicast group. The values shown in Table A.2 have been assigned.

Frequently used permanently assigned multicast addresses include FF02::1 for all nodes and FF02::2 for all routers; both have link-local scope and are therefore never forwarded by a router.

A.3   IPv6 Neighbor Discovery

The IPv6 Neighbor Discovery (ND) protocol is documented in [RFC4861]. As discussed in more detail in Section 3.2, ND is used for managing the relationship of IPv6 nodes to

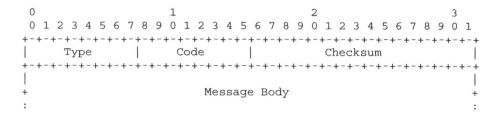

Figure A.7 General format of ICMPv6 messages.

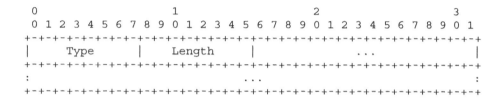

Figure A.8 General format of an ICMPv6 message option.

their neighbors on the same link. This appendix shows a few packet formats and protocol exchanges of the ND base protocol.

All ND messages are messages of the ICMPv6 protocol [RFC4443], which directly follows the IPv6 and extension headers (i.e. no TCP or UDP header is involved), and have the general format shown in Figure A.7.

The type field determines the message format and its specific meaning; the code field can be used to distinguish subtypes, but is always set to zero by the ND base protocol. The checksum field is an Internet checksum (16-bit one's complement of the one's complement sum of the entire ICMPv6 message, prepended with the usual pseudo-header constructed from some IPv6 header fields [RFC2460, section 8.1]).

Certain ND messages can include *options*, each of which is encoded as shown in Figure A.8. The type field identifies a specific kind of option, and the length gives the length of the option in multiples of eight bytes (i.e. options are always padded to 64-bit boundaries; the length of 0 is not allowed). Options of certain types can appear more than once in an ND message.

One function of the ND protocol is for hosts to establish communication with routers and to obtain the salient parameters of the link they are on.

On each of its multicast-capable interfaces, each router periodically multicasts a *Router Advertisement* (RA) message. Hosts use these messages to obtain a list of candidate routers to send off-link packets to.

Router Advertisement messages are also used to inform hosts about certain parameters to be used by hosts on the link. In particular, a value that hosts should use in outgoing packets

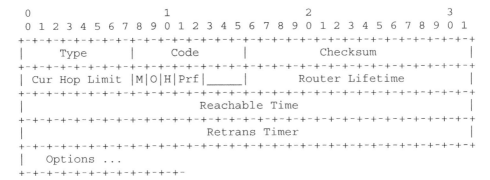

Figure A.9 IPv6 Router Advertisement message.

for the IPv6 Hop Limit field is part of the base format. There are also a number of bits and short fields:

M (managed) If set, IPv6 addresses are to be obtained from a DHCPv6 server [RFC3315] instead of using Stateless Address Autoconfiguration (SAA, see Appendix A.4). (Note that the meaning of this bit is redefined by 6LoWPAN-ND.)

O (other) If set and the M bit is not set, only additional configuration information such as DNS server addresses is to be obtained via DHCPv6.

H (Home Agent) Set by mobile-IP Home Agents.

Prf (preference) A two-bit signed integer indicating how much to prefer the advertising router over other potential default routers; only the values $+1$ (high), 0 (default) and -1 (low) are defined.

Options can be used to set link parameters such as the link maximum transmission unit (MTU) and the prefixes available on the link (see the prefix information option below). There is even an option (currently experimental) to specify addresses of DNS servers in Router Advertisements [RFC5006]. The assumption here is that routers need to undergo some form of configuration anyway and the Router Advertisements can then be used to automatically transfer configuration information to the hosts with respect to these parameters.

The Router Advertisement format, defined in [RFC4861] and modified by [RFC3775] and [RFC4191], is illustrated in Figure A.9.

Hosts may wait for the periodic Router Advertisement messages sent by each router to the all-nodes multicast address (FF02::1), or they may prompt routers to send RAs more quickly by sending out a *Router Solicitation* (RS) to the all-routers multicast address (FF02::2). The base part of the RS format is essentially an empty message, see Figure A.10.

Router Advertisement (RA) messages may contain *Prefix Information Options*, see Figure A.11. Each option specifies a prefix (initial bits of a unicast address) and its length, as well as whether this prefix is applicable for Stateless Address Autoconfiguration and for on-link decisions.

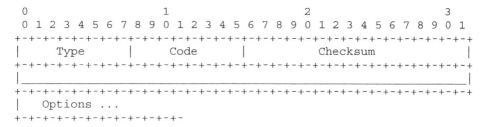

Figure A.10 IPv6 Router Solicitation message.

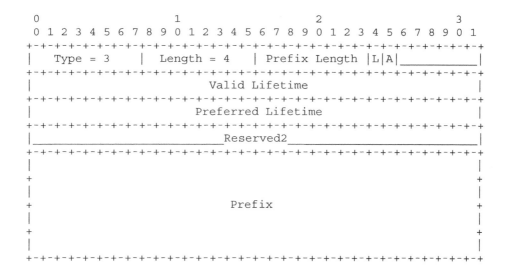

Figure A.11 IPv6 ND prefix information option.

The *prefix length* gives the number of bits that are valid in the *prefix* field (i.e. a prefix length of 64 indicates that the prefix is a /64); note that all 128 bits of the prefix field are always included. The L bit, if set, indicates that the prefix given can be used for on-link determination, i.e. all addresses covered by the prefix are to be considered on-link on the interface where this RA has been received. If it is not set, the option neither denies nor confirms anything about on-link determination. The A bit, if set, indicates that this prefix is meant to be used for Stateless Address Autoconfiguration – see Appendix A.4.

Other ND messages not further discussed in this quick refresher include:

Neighbor solicitation messages are sent by a node to find out the link-layer address of another on-link node, or to ascertain that an on-link node is still reachable via a cached link-layer address (*Neighbor Unreachability Detection*, NUD). Neighbor solicitations are also used in the process of Duplicate Address Detection (DAD) – see Section A.4 below.

Neighbor advertisement messages are responses to Neighbor Solicitation messages (or could be sent unsolicited to announce a link-layer address change).

Redirect messages are sent by a router that forwarded a packet for a host to inform it of a better first hop towards the packet's destination, including to inform the host that the destination is in fact on-link.

A.4 IPv6 Stateless Address Autoconfiguration

IPv6 Stateless Address Autoconfiguration (SAA) is documented in [RFC4862], with privacy extensions defined in [RFC4941]. SAA has been defined to enable automatic configuration of host addresses in cases where the specific address used matters less than its uniqueness and the ability to properly route packets to this address.

SAA is based on the concept of an *interface identifier* (IID) that is derived in a link-dependent way for each interface on the link, ensuring uniqueness at least per link [RFC4291]. For Ethernet, this is usually a modified EUI-64 computed from the MAC-48 by inserting FF-FE in the middle and then flipping the universal/local bit in the EUI-64 to form the IID. This IID is combined with a prefix to form an address. Addresses formed this way go through three stages:

tentative – the uniqueness of the address has not yet been verified. The address is not yet actually being used except for Neighbor Discovery packets that perform Duplicate Address Detection for the tentative address.

preferred – the address is in use and active. Nodes obtain addresses for their interfaces with a certain lifetime (which may be infinite or a finite number of seconds), so that an address can possibly be reassigned later to a different node or interface. The *preferred lifetime* is the time that the address stays *preferred*, after which the address becomes *deprecated*.

deprecated – the address is still *valid* but on its way out. New communications should no longer use it as a source address. Existing communications may continue to use it if it would be too hard to switch over to a new address, e.g. an existing TCP connection may continue to operate on a deprecated address until the *valid lifetime* finally expires. The difference between the preferred and the valid lifetime provides for a grace period in switching over to a new address.

While an IID is supposed to be unique on a link, a second safety net is provided by the *Duplicate Address Detection* (DAD) performed on a tentative address before it becomes preferred. In standard IPv6, DAD is meant to be run before commencing use of any kind of address, even addresses assigned via DHCPv6 or addresses configured into routers.

For nodes that frequently find themselves on new links, the delay implied by performing DAD before communication can continue can be prohibitive. [RFC4429] therefore introduces *optimistic DAD*, which adds another state an address can be in:

optimistic – the address is in the process of being validated by DAD, but can already be used for certain kinds of communication. The effects of this state are similar to those of the deprecated state, except that the node will also avoid any action that would force the not yet validated address into other nodes' neighbor caches. Actually, optimistic is not entirely a separate state, but a temporary restriction that can apply to addresses that, depending on their lifetimes, change to preferred or deprecated state once validated.

DAD requires the use of link-wide multicast to search out other users of what is still a tentative (or optimistic) address for the current node. As that would be rather expensive and only of limited assurance in a multihop network with sleeping nodes, 6LoWPAN-ND replaces DAD by a different mechanism. (Note that the SAA specification explicitly allows this when the variable *DupAddrDetectTransmits* is set to zero [RFC4862, Section 5.4].)

The second element of SAA is the dissemination of appropriate prefixes to be used with the locally built IIDs for address creation. SAA uses prefix information options in Router Advertisements (RAs) to obtain these prefixes, along with values for the preferred and valid lifetimes of the addresses created. Later RAs may extend these lifetimes, may count them down or actively reduce them to expiry or may simply stop advertising the prefix at all, leaving the nodes to count down the time.

B

IEEE 802.15.4 Reference

B.1 Introduction

The IEEE 802.15.4 standards define low-power wireless radio techniques for what the IEEE calls wireless personal area networks (WPANs). Today IEEE 802.15.4 is a vastly popular radio standard used with a much broader range of applications than what the term WPAN might seem to describe. It is aimed at providing cheap, low-power, short-range communications for embedded devices. Several other standards or stack specifications use IEEE 802.15.4 as their physical layer (PHY) and data link-layer (DLL), including 6LoWPAN, ISA100 and ZigBee specifications.

The latest version of the standard is IEEE 802.15.4-2006 [IEEE802.15.4]. Channel sharing is achieved using carrier sense multiple access (CSMA), and acknowledgments are provided for reliability. Link-layer security is provided with 128-bit AES encryption, described further in Appendix B.3. Addressing modes utilizing 64-bit and 16-bit addresses are provided with unicast and broadcast capabilities. The payload of the physical frame can be up to 127 bytes in size, with 72–116 bytes of payload available after link-layer framing, depending on a number of addressing and security options.

The MAC can be run in two modes: beaconless and beacon-enabled. Beaconless mode uses a pure CSMA channel access and operates quite like basic IEEE 802.11 without channel reservations. Beacon-enabled mode is more complex, with a superframe structure and the possibility to reserve time-slots for critical data. IEEE 802.15.4 includes a large range of mechanisms for forming networks and superframe control.

The following frame types are defined by the standard:

Data frames for the transport of actual data, discussed further in Appendix B.2;

Acknowledgment frames that are meant to be sent back by a receiver immediately after successful reception of a data frame, if requested by the acknowledgment request bit in the data frame MAC header;

Table B.1 Frequency ranges and channels for IEEE 802.15.4.

Frequency range (MHz)	Region	Channel numbers	Bit rate (kbit/s)
868	Europe	0	20
902–928	US	1–10	40
2400–2483.5	worldwide	11–26	250

MAC layer command frames, used mainly in beacon-enabled mode to enable various MAC layer services such as association to and disassociation from a coordinator, and management of synchronized transmission; and

Beacon frames, used by a coordinator in beacon-enabled mode to structure the communication with its associated nodes.

IEEE 802.15.4 radios are available for three frequency ranges, two of which are region-specific and one available worldwide. Depending on the frequency range employed, data rates vary from 20 to 250 kbit/s. These frequency ranges are partitioned into *channels* which are identified by channel numbers – see Table B.1 [IEEE802.15.4, section 6.1.2]. (Additional channel numbers are defined to specify alternative modulation formats.)

B.2 Overall Packet Format

IEEE 802.15.4 provides for a number of different MAC layer packet formats. The most important one for 6LoWPAN is the actual data packet. The PHY and MAC layer parts of this packet format are shown in Figure B.1. If you are used to IEEE 802.15.4, this figure may look unusual, as it is designed with the conventions in mind that are used in IETF RFC box diagrams (see also Appendix A.1): the figure uses a most-significant-byte-first byte order as well as a most-significant-bit-first bit order within the bytes. Note that, in the actual air interface, the PHY layer header is preceded by a preamble and a starting delimiter; as these are specific to the variant of IEEE 802.15.4 in use, they are not shown here.

The data packet starts with a one-byte physical layer header that contains a one-bit reserved field (indicated here with an underscore) and a seven-bit length field, which counts the entire length of the remaining packet including the trailing checksum in bytes. Given its field length, this means the count can be between 0 and 127; the overall data packet (including the physical header) can therefore be between 1 and 128 bytes long. Three bytes are used for the fixed part of the MAC layer header, followed by variable parts, the presence and length of which depend on settings in the fixed part. The actual payload follows (possibly encrypted and possibly including 4, 8, or 16 additional bytes of message integrity check), followed by a 16-bit CRC (cyclic redundancy check) value used as the frame check sequence (FCS). A synopsis for the meaning of the various fields in the fixed MAC header is provided in Table B.2.

The contents of the variable header part labeled "Addresses" in Figure B.1 depends on the values of the SAM, DAM and C fields. The SAM and DAM fields select between a zero-length address (addressing the PAN coordinator), a 16-bit address or a 64-bit address. For the latter two cases, the address is preceded by a 16-bit PAN identifier, except if the C bit is set:

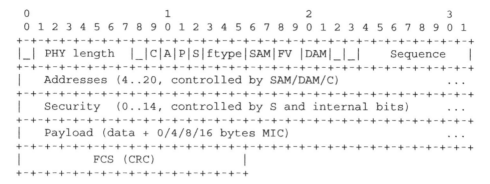

Figure B.1 Overall structure of the IEEE 802.15.4 data packet.

Table B.2 Fields in the fixed header/trailer of the IEEE 802.15.4 data packet.

–	(reserved fields)
C	PAN ID compression
A	ACK request
P	Frame pending
S	Security enabled
ftype	Frame type (001 binary for data packets)
SAM	Source addressing mode (see Table B.3)
FV	Frame version (00 binary for compatibility with 2003 version, 01 binary for frames only compatible with 2006 version)
DAM	Destination addressing mode (see Table B.3)
Sequence	Sequence number (for ACK)
FCS	Frame check sequence

Table B.3 Addressing modes in the IEEE 802.15.4 MAC header.

00	Neither PAN identifier nor the address field is given
01	Reserved
10	Address field contains a 16-bit short address
11	Address field contains a 64-bit extended address

this indicates that the two PAN identifiers are the same and that it therefore was elided on the source address field (SAM and DAM cannot be both zero, i.e. at least one address must be given). Depending on the values of these three fields, the total length of the "Addresses" part can be between 4 and 20 bytes.

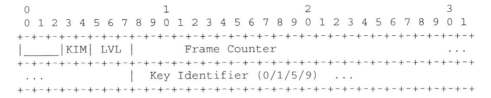

```
 0                   1                   2                   3
 0 1 2 3 4 5 6 7 8 9 0 1 2 3 4 5 6 7 8 9 0 1 2 3 4 5 6 7 8 9 0 1
+-+-+-+-+-+-+-+-+-+-+-+-+-+-+-+-+-+-+-+-+-+-+-+-+-+-+-+-+-+-+-+-+
|_____|KIM| LVL |          Frame Counter                   ...
+-+-+-+-+-+-+-+-+-+-+-+-+-+-+-+-+-+-+-+-+-+-+-+-+-+-+-+-+-+-+-+-+
 ...              |  Key Identifier (0/1/5/9)  ...
+-+-+-+-+-+-+-+-+-+-+-+-+-+-+-+-+-+-+-+-+-+-+-+-+-+-+-+-+-+-+-+-+
```

Figure B.2 The security subheader in an IEEE 802.15.4 data packet.

Table B.4 Key identifier modes (KIM) in the IEEE 802.15.4 security subheader.

00	Key identified by source and destination
01	Key identified by macDefaultKeySource + one-byte key index
10	Key identified by four-byte key source + one-byte key index
11	Key identified by eight-byte key source + one-byte key index

Table B.5 Security level (LVL) in the IEEE 802.15.4 security subheader.

000	No security
001	4-byte MIC
010	8-byte MIC
011	16-byte MIC
100	Encryption only
101	Encryption + 4-byte MIC
110	Encryption + 8-byte MIC
111	Encryption + 16-byte MIC

B.3 MAC-layer Security

Section 3.3.2 discusses link-layer security in general. This appendix lists the most important formats employed by IEEE 802.15.4 to perform its security functions.

The S bit in the data packet base header controls whether the packet contains a security subheader, which, if present, looks as illustrated in Figure B.2.

The key identifier mode (KIM) field in the security subheader specifies the structure of the key identifier field as listed in Table B.4.

Finally, the security level (LVL) field in the security subheader specifies which security functions are employed and, if a message integrity check (MIC) field is used, to how many bytes that is truncated – see Table B.5.

List of Abbreviations

6LoWPAN IPv6 over low-power wireless area networks

AAL ATM adaptation layer

ACK acknowledgement

AH authentication header

AES advanced encryption standard

AMI advanced metering infrastructure

ANSI American National Standards Institute

ANubis Advanced Network for Unified Building Integration and Services

ASHRAE American Society of Heating Refrigeration and Air-conditioning Engineers

AODV ad hoc on-demand distance vector

API application programming interface

ATM asynchronous transfer mode

AVP (RTP) audio video profile

BACnet building automation and control networks

BER binary encoding rules

BGP border gateway protocol

BLIP Berkeley IP implementation

BXML binary XML

BVLL BACnet virtual link layer

CAP compact application protocol

CBC cipher-block chaining

CCM counter with CBC-MAC

6LoWPAN: The Wireless Embedded Internet   Zach Shelby and Carsten Bormann
© 2009 John Wiley & Sons, Ltd

CID Context ID

CoA care-of address

COSEM companion specification for energy metering

CPU central processing unit

CRC cyclical redundancy check

CSMA carrier sense multiple access

DAD Duplicate Address Detection

DHCP dynamic host configuration protocol

DLL data link layer

DLMS device language message specification

DNS domain name system

DPWS devices profile for web services

DSCP differentiated services control point

DSL digital subscriber line

DSR dynamic source routing

DYMO dynamic MANET on-demand

EAP extensible authentication protocol

ECN explicit congestion notification

EPSEM extended protocol specification for electronic metering

ER edge router

ESP encapsulating security payload

ETSI European Telecommunications Standards Institute

EU European Union

EUI extended unique identifier

EXI efficient XML interchange

FFD full-function device

FIB forwarding information base

FIND future Internet design

FCS frame check sequence

FTP file transfer protocol

GENA general event notification architecture

GMT Greenwich mean time

GPRS general packet radio system

GSM global system for mobile communications

GTK group transient key

GTS guaranteed time slot

HA Home Agent

HC header compression

HTML hypertext markup language

HTTP hypertext transfer protocol

HVAC heating, ventilating, and air-conditioning

IANA Internet Assigned Numbers Authority

ICV integrity check value

ID Internet draft

IEEE Institute of Electrical and Electronics Engineers

IETF Internet Engineering Task Force

IID interface identifier

IKE internet key exchange

IMS IP multimedia subsystem

IP Internet Protocol

IPv4 Internet Protocol version 4

IPv6 Internet Protocol version 6

IPsec Internet Protocol security

IPSO IP for Smart Objects (Alliance)

ISA International Society of Automation (formerly Instrument Society of America)

ISDN integrated services digital network

ISM industrial, scientific and medical

ISA International Organization for Standardization

ISP Internet service provider

IT information technology

KNX Konnex (protocol)

L1 Layer 1 (physical layer)

L2 Layer 2 (link layer)

L3 Layer 3 (network layer)

L4 Layer 4 (transport layer)

L7 Layer 7 (application layer)

LAN local area network

LBR LLN border router

LLN low-power and lossy network

LMA local mobility anchor

LoWPAN low-power wireless area network

M2M machine-to-machine

MAC message authentication code, or

MAC medium access control

MAG mobile access gateway

MANEMO mobile ad hoc network mobility

MANET mobile ad hoc network

MIB management information base

MIC message integrity check

MIME multipurpose Internet mail extensions

MIP Mobile IP

MIPv6 Mobile IP version 6

MNN mobile network node

MPR multipoint relay

MQTT MQ telemetry transport

MQTT-S MQ telemetry transport for sensor

MTR multi-topology routing

MTU maximum transmission unit

NA Neighbor Advertisement

NALP not a lowpan packet

NanoWS Nano web services

NAT network address translator

NC Node Confirmation

ND Neighbor Discovery

NEMO network mobility

NETLMM network-based local mobility management

NR Node Registration

NS Neighbor Solicitation

NUD Neighbor Unreachability Detection

OASIS Organization for the Advancement of Structured Information Standards

oBIX Open Building Information Exchange

OGC Open Geospatial Consortium

OLSR optimized link-state routing

OS operating system

OSI open systems interconnection

OSPF open shortest path first

OUI organizationally unique identifier

PAN personal area network

PC personal computer

PDU protocol data unit

PHA proxy Home Agent

PHY physical layer

PILC performance implications of link characteristics (former IETF WG)

PLC power line communications

PLPMTUD packetization layer path MTU discovery

PMIPv6 proxy Mobile IPv6

PMK pairwise master key

PMTUD path MTU discovery

PPP point-to-point protocol

PSEM protocol specification for electric metering

PSK pre-shared key

PTK pairwise transient key

QoS quality of service

RA Router Advertisement

RERR route error

REST representational state transfer

RDF resource description framework

RF radio frequency

RFD reduced-function device

RFC request for comments

RFID radio frequency identification

RIB routing information base

RIP routing information protocol

ROHC robust header compression (IETF WG and suite of standards)

ROLL routing over low-power and lossy networks (IETF WG)

RPC remote procedure call

RREP route reply

RREQ route request

RS Router Solicitation

RSSI received signal strength indicator

RTP real-time transport protocol

RTCP RTP control protocol

SAA Stateless Address Autoconfiguration

SAR segmentation and reassembly

SCADA supervisory control and data acquisition

SCTP stream control transmission protocol

SDP session description protocol

SICS Swedish Institute for Computer Science

SIP session initiation protocol

SLP service location protocol

SNAP simple network access protocol, or

SNAP subnetwork access protocol

SNMP simple network management protocol

SOAP simple object access protocol

SoC system on a chip

SPI serial peripheral interface

SSDP simple service discovery protocol

SSLP simple service location protocol

SSID service set identifier

TBRPF topology dissemination based on reverse-path forwarding

TCP transmission control protocol

TCO total cost of ownership

TDMA time division multiple access

TID Transaction Identifier

TTL Time to Live

UART universal asynchronous receiver/transmitter

UDP user datagram protocol

ULA unique local (unicast) address

UMTS universal mobile telecommunications system

URL uniform resource locator

URI uniform resource identifier

USB universal serial bus

UPnP universal plug-and-play

UUID universally unique identifier

VoIP voice-over-IP

VPN virtual private network

W3C World Wide Web Consortium

WADL web application description language

WAP wireless application protocol

WBXML WAP binary XML

WDS wireless distribution system

WEP wired equivalent privacy

WEI Wireless Embedded Internet

WG working group

WLAN wireless local area network

WPA wireless protected access

WPAN wireless personal area network

WPC watt pulse communication

WSDL web services description language

WSN wireless sensor network

WS web service

WWW World Wide Web

XML extensible markup language

XMPP extensible messaging and presence protocol

ZAL ZigBee application layer

ZCL ZigBee cluster library

Glossary

Address resolution The process of determining the corresponding link-layer address for an IP address on a link.

Ad hoc LoWPAN An isolated LoWPAN, not connected to any other IP networks. Ad hoc LoWPANs make use of unique local IPv6 unicast addresses (ULAs).

Anycast Send a packet from one interface to just one out of a defined set of interfaces.

Anycast address An IP address used by a set of interfaces, usually belonging to different nodes. A packet sent to an anycast address is sent to the nearest interface in that set by a routing protocol. An anycast address is formatted in an identical manner to a unicast address. See Section A.2 for more information on IPv6 addressing.

Availability The property that the system is available for use as intended, more specifically the security objective that the system is not subject to *denial of service* attacks. See Section 3.3.1.

Backbone link An IPv6 link that interconnects two or more LoWPAN Edge Routers in an Extended LoWPAN topology, two or more LLN border routers in the ROLL architecture or two or more backbone routers in ISA100.

Bootstrapping During setup of a node, the establishment of state required for operation that can be performed automatically, without human intervention. With respect to the network configuration, performed by 6LoWPAN-ND, see Section 3.2.

Border routing Routing between two different routing domains. In 6LoWPAN, border routing is performed either by the LoWPAN Edge Router or by a router on the backbone link. Border routing is covered in Section 4.2.7.

Care-of address, CoA An IP address acquired by a mobile node when roaming to a visited network in Mobile IP.

Commissioning During initial setup of a node, the establishment of state required for operation that requires human intervention – see Section 3.1.

Confidentiality The security objective that data is not made available to unintended parties, more specifically, cannot be overheard by unintended listeners – see Section 3.3.1.

Coordinated universal time See UTC below.

6LoWPAN: The Wireless Embedded Internet   Zach Shelby and Carsten Bormann
© 2009 John Wiley & Sons, Ltd

Correspondent node A node which a mobile node communicates with in Mobile IP.

Distance-vector routing Routing algorithms based on variations of the Bellman–Ford algorithm. Using this approach, each link (and possibly node) is assigned a cost using appropriate route metrics. When sending a packet from node A to node B, the path with the lowest cost is chosen. The routing table of each router keeps soft-state route entries for the destinations it knows about, with the associated path cost. Routing information is updated either proactively (a priori) or reactively (on-demand) depending on the routing algorithm.

Duplicate address detection, DAD Automatic detection of configuration errors that lead to interfaces of multiple nodes on a link having the same address. Part of standard Neighbor Discovery (ND). See Appendix A.4, and Section 3.2.2 for how the same effect is achieved in 6LoWPAN.

Extended LoWPAN The aggregation of multiple LoWPANs interconnected by a backbone link via edge routers and forming a single subnet.

Handover A process in which a mobile node disconnects from its existing point of attachment and attaches itself to a new point of attachment. Handover may include operations at specific link layers, as well as at the IP layer, that enable the mobile node to communicate again. One or more application streams typically accompany the mobile node as it undergoes handover.

Home address The IP address of a mobile node on its home network in Mobile IP.

Home agent A Mobile IP router on the home network of a mobile node which performs forwarding for it while the mobile node is roaming.

Home network The network that a mobile node belongs to when not roaming in Mobile IP.

Integrity The property that data stay as intended; more specifically, the security objective that data cannot be altered by unauthorized parties – see Section 3.3.1.

Link A communication facility of medium over which IP nodes can communicate at the link layer (the layer just below IP). In 6LoWPAN, *link-local scope* refers to communication using a single link-layer transmission. *Local* refers to an address of a destination considered to be within link-local scope. *Non-local* is the opposite of local.

Link-state routing Routing algorithm where each node acquires complete information about the entire network, often called a graph. To do this each node floods the network with information about its links to nearby destinations. After receiving link-state reports from sufficient nodes, each node then calculates a tree with the shortest path (least cost) from itself to each destination using e.g. Dijkstra's algorithm. This tree is used either to maintain the routing table in each node for hop-by-hop forwarding, or to include a source-route in the header of the IP packet.

LoWPAN Edge Router An IPv6 router that interconnects the LoWPAN to another IP network.

LoWPAN host A node that only sources or sinks IPv6 datagrams.

LoWPAN Node A node that composes a LoWPAN, referring to both hosts and routers.

LoWPAN Router A node that forwards datagrams between arbitrary source–destination pairs using a single 6LoWPAN interface performing IP routing on that interface.

IPsec IP security, the standard security architecture for the Internet Protocol – see Section 3.3.3.

Inter-domain routing Routing between different administrative domains. The border gateway protocol (BGP) is used for inter-domain routing on the core Internet.

Interface identifier, IID The part of an IPv6 address used to identify an interface on a link, which must be unique in a subnet. In 6LoWPAN the interface identifier (IID) is created from the link-layer address of the interface.

International atomic time See TAI below.

Internet key exchange, IKE Protocol used with IPsec to dynamically establish a security association between Internet nodes.

Intra-domain routing Routing within the same administrative domain. OSPF and AODV are examples of intra-domain routing protocols.

Macro-mobility Refers to mobility between networks. In 6LoWPAN we consider macro-mobility to refer to mobility between LoWPANs, such that the IPv6 prefix changes.

Maximum transmission unit, MTU The maximum size of a packet that can be carried in one transmission unit over a link.

Medium access control, MAC A sublayer of the data link layer (DLL) which is responsible for addressing and channel access to a shared medium.

Mesh-Under Refers to multihop forwarding in a LoWPAN using link-layer techniques.

Micro-mobility Refers to mobility that occurs within a network domain. In 6LoWPAN we can consider micro-mobility to refer to the mobility of a node within a LoWPAN where the IPv6 prefix does not change.

Multicast Send a packet from one interface to every one of a defined set of interfaces in one go.

Multicast address An IP address used by a set of interfaces, usually belonging to different nodes. A packet sent to a multicast address is delivered to all interfaces identified by that address (unless that is prevented by packet loss). See Section A.2 for more information on IPv6 addressing.

Neighbor Discovery, ND A protocol used by IPv6 nodes to perform operations between nodes on a link such as address resolution, duplicate address resolution and Neighbor Unreachability Detection. See Section A.3 for a detailed description.

Neighbor unreachability detection, NUD Automatic detection that a neighbor is no longer reachable in the way previously used. Part of standard Neighbor Discovery (ND). NUD may trigger recovery action such as switching to a different default router. See Section 3.2.5 for the way the same effect is achieved in a LoWPAN.

Network mobility Mobility in which an entire IP network moves its point of attachment.

Node mobility Mobility in which an IP node moves between points of attachment.

Prefix The first bits of an IP address common to addresses using it. An IPv6 prefix has an associated prefix length, i.e. the number of those bits.

Proactive routing Routing algorithms using a proactive approach build up routing information on node before the routes are needed. Thus they proactively prepare for the data traffic by learning routes to all possible or likely destinations.

Reactive routing Reactive routing protocols store little or no routing information after autoconfiguration of the routing algorithm. Instead, routes are discovered dynamically only at the time they are needed. Thus a process called route discovery is executed when a router receives a packet to an unknown destination.

Roaming A process in which a mobile node moves from one network to another, typically with no existing packet streams.

Route-Over Refers to multihop forwarding in a LoWPAN using IP routing.

Routing table A table where routers keep entries with next hop information.

Simple LoWPAN A Simple LoWPAN consists of a single edge router and the set of LoWPAN Nodes on the same subnet.

Stateless address autoconfiguration, SAA An ND technique which enables automatic configuration of host addresses by deriving the interface identifier using link-layer information, usually the EUI-64 of the interface.

Subnet A subnet refers to a group of nodes that have the same IP prefix. All nodes in a subnet are considered to be on the same link. In 6LoWPAN multiple IP hops may be required to connect all nodes on a link because of properties of the wireless link.

TAI International atomic time, an internationally maintained standard monotonic timescale based on the most precise clocks available. The civil time UTC is derived from TAI by occasionally inserting leap seconds to compensate for the irregular rotation of the earth; as of January 2009, UTC is 34 seconds behind TAI. As the progress of TAI is not disturbed by leap seconds, it is more appropriate than UTC for applications such as process control.

Unicast Send a packet from one interface to one other interface.

Unicast address An IP address assigned to a single interface. See Section A.2 for more information on IPv6 addressing.

Uniform resource identifier, URI A compact sequence of characters that identifies an abstract or physical resource, with a structure that contains a URI scheme (such as "http") as well as certain scheme-dependent elements such as a network location or authority, a path, a query, and/or a fragment identifier. A subset of URIs, the uniform resource locator (URL) provides a means of locating the resource by describing how to access it at a location in the network. The terms URI and URL are rarely distinguished carefully.

UTC Coordinated Universal Time, the basis for the civil timescales used throughout the world, which are modified from UTC by the locally applicable timezone. UTC is based on TAI, but synchronized to the rotation of the earth by the occasional insertion of leap seconds. Often still colloquially designated with the no longer correct name of its historic precursor GMT, Greenwich mean time.

Visited network The network that a mobile node is visiting in Mobile IP when roaming.

Web services A client–server communication paradigm utilizing HTTP for machine-to-machine interaction on the web. See Section 5.3.4 for an introduction to web services and Section 5.4.1 for application to 6LoWPAN.

Whiteboard A conceptual data structure similar to a MIPv6 binding cache which may be supported by edge routers. The whiteboard is used for performing DAD and NUD across the entire LoWPAN. The whiteboard contains bindings for LoWPAN Nodes consisting of owner interface identifier, IPv6 address, timeout, along with transaction ID history – see Section 3.2.2.

References

[2002/91/EC] European Union. Directive 2002/91/EC On the Energy Performance of Buildings, December 2002.

[4WARD] EU FP7 4WARD Project. http://www.ict-forward.eu.

[6LoWPAN] IETF 6LoWPAN Working Group. http://tools.ietf.org/wg/6lowpan.

[AES] Specification for the Advanced Encryption Standard (AES). Federal information processing standards publication 197, 2001.

[ANSI] American National Standards Institute. http://www.ansi.org.

[BACnet] American Society of Heating Refrigeration and Air-Conditioning Engineers. BACnet – A Data Communication Protocol for Building Automation and Control Networks (ASHRAE/ANSI 135-2008). Tech. Rep. ISBN/ISSN: 1041-2336, ASHRAE, 2008.

[Baden06] Baden, S., Fairey, P., Waide, P., de T'serclaes, P. and Laustsen, J. Hurdling Financial Barriers to Low Energy Buildings: Experiences from the USA and Europe on Financial Incentives and Monetizing Building Energy Savings in Private Investment Decisions. In *ACEEE Summer Study on Energy Efficiency in Buildings*. American Council for an Energy Efficient Economy, August 2006.

[Bauge08] Bauge, T., Gluhak, A., Presser, M. and Herault, L. Architecture Design Considerations for the Evolution of Sensing and Actuation Infrastructures in the Future Internet. In *WPMC*, 2008.

[BLIP] The Berkeley IP Implementation. http://smote.cs.berkeley.edu:8000/tracenv/wiki/blip.

[BXML] Open Geospatial Consortium. Binary Extensible Markup Language (BXML) Encoding Specification. Tech. rep., 03-002r9, 2006.

[CC1101] Low-Cost, Low-Power Sub-1 GHz RF Transceiver CC1101, Texas Instruments. http://focus.ti.com/lit/ds/symlink/cc1101.pdf.

[Contiki] The Contiki Operating System. http://www.sics.se/contiki.

[DEFRA] UK Department for Environment, Food and Rural Affairs. http://www.defra.gov.uk.

[DLMS] DLMS Association. http://www.dlms.com.

[DoE06] US Department of Energy. Annual Energy Review, June 2007.

[Dolev81] Dolev, D. and Yao, A. On the security of public key protocols. In *Proceedings of the IEEE 22nd Annual Symposium on Foundations of Computer Science*, pp. 350–357. 1981.

[DPWS] Devices Profile for Web Services Version 1.1. http://docs.oasis-open.org/ws-dd/dpws/wsdd-dpws-1.1-spec.html. Tech. rep., OASIS, July 2009.

[Dunkels03] Dunkels, A. Full TCP/IP for 8-Bit Architectures. In *Proceedings of the First ACM/Usenix International Conference on Mobile Systems, Applications and Services (MobiSys 2003)*. ACM, San Francisco, May 2003.

[ETSI] European Telecommunications Standards Institute. http://www.etsi.org.

[EUI-64] Guidelines For 64-Bit Global Identifier (EUI-64) Registration Authority. http://standards.ieee.org/regauth/oui/tutorials/eui64.html.

[EXI] Efficient XML Interchange (EXI) Primer. http://www.w3.org/TR/exi-primer.

[FI] Fast Infoset. http://en.wikipedia.org/wiki/FastInfoset.

[FIAssembly] European Future Internet Assembly. http://www.future-internet.eu.

[FIND] NSF Future Internet Design. http://www.nets-find.net.

[ID-6lowpan-hc] Hui, J. and Thubert, P. Compression Format for IPv6 Datagrams in 6LoWPAN Networks. Internet-Draft draft-ietf-6lowpan-hc-05, Internet Engineering Task Force, Jun. 2009. Work in progress.

[ID-6lowpan-mib] Kim, K., Mukhtar, H., Yoo, S. and Park, S.D. 6LoWPAN Management Information Base. Internet-Draft draft-daniel-6lowpan-mib-00, Internet Engineering Task Force, March 2009. Work in progress.

[ID-6lowpan-mipv6] Silva, R. and Silva, J. An Adaptation Model for Mobile IPv6 support in LoWPANs. Internet-Draft draft-silva-6lowpan-mipv6-00, Internet Engineering Task Force, May 2009. Work in progress.

[ID-6lowpan-nd] Shelby, Z., Thubert, P., Hui, J., Chakrabarti, S., Bormann, C. and Nordmark, E. 6LoWPAN Neighbor Discovery. Internet-Draft draft-ietf-6lowpan-nd-06, Internet Engineering Task Force, Sep. 2009. Work in progress.

[ID-6lowpan-rr] Kim, E., Kaspar, D., Gomez, C. and Bormann, C. Problem Statement and Requirements for 6LoWPAN Routing. Internet-Draft draft-ietf-6lowpan-routing-requirements-04, Internet Engineering Task Force, Jul. 2009. Work in progress.

[ID-6lowpan-sslp] Kim, K., Yoo, S., Lee, H., Park, S.D. and Lee, J. Simple Service Location Protocol (SSLP) for 6LoWPAN. Internet-Draft draft-ietf-6lowpan-sslp-01, Internet Engineering Task Force, 2007. Work in progress.

[ID-6lowpan-uc] Kim, E., Kaspar, D., Chevrollier, N. and Vasseur, J. Design and Application Spaces for 6LoWPANs. Internet-Draft draft-ietf-6lowpan-usecases-03, Internet Engineering Task Force, Jul. 2009. Work in progress.

[ID-despres-6rd] Despres, R. IPv6 Rapid Deployment on IPv4 infrastructures (6rd). Internet-Draft draft-despres-6rd-03, Internet Engineering Task Force, Apr. 2009. Work in progress.

[ID-global-haha] Thubert, P., Wakikawa, R. and Devarapalli, V. Global HA to HA protocol. Internet-Draft draft-thubert-mext-global-haha-00, Internet Engineering Task Force, Mar. 2008. Work in progress.

[ID-manet-dymo] Chakeres, I. and Perkins, C. Dynamic MANET On-demand (DYMO) Routing. Internet-Draft draft-ietf-manet-dymo-17, Internet Engineering Task Force, Mar. 2009. Work in progress.

[ID-manet-nhdp] Clausen, T., Dearlove, C. and Dean, J. MANET Neighborhood Discovery Protocol (NHDP). Internet-Draft draft-ietf-manet-nhdp-09, Internet Engineering Task Force, Mar. 2009. Work in progress.

[ID-manet-olsrv2] Clausen, T., Dearlove, C. and Jacquet, P. The Optimized Link State Routing Protocol version 2. Internet-Draft draft-ietf-manet-olsrv2-08, Internet Engineering Task Force, Mar. 2009. Work in progress.

[ID-nemo-pd] Droms, R., Thubert, P., Dupont, F. and Haddad, W. DHCPv6 Prefix Delegation for NEMO. Internet-Draft draft-ietf-mext-nemo-pd-02, Internet Engineering Task Force, Mar. 2009. Work in progress.

[ID-roll-building] Martocci, J., Riou, N., Mil, P. and Vermeylen, W. Building Automation Routing Requirements in Low Power and Lossy Networks. Internet-Draft draft-ietf-roll-building-routing-reqs-07, Internet Engineering Task Force, Sep. 2009. Work in progress.

[ID-roll-fundamentals] Thubert, P., Watteyne, T., Shelby, Z. and Barthel, D. LLN Routing Fundamentals. Internet-Draft draft-thubert-roll-fundamentals-01, Internet Engineering Task Force, Apr. 2009. Work in progress.

[ID-roll-home] Brandt, A., Buron, J. and Porcu, G. Home Automation Routing Requirements in Low Power and Lossy Networks. Internet-Draft draft-ietf-roll-home-routing-reqs-08, Internet Engineering Task Force, Sep. 2009. Work in progress.

[ID-roll-indus] Networks, D., Thubert, P., Dwars, S. and Phinney, T. Industrial Routing Requirements in Low Power and Lossy Networks. Internet-Draft draft-ietf-roll-indus-routing-reqs-06, Internet Engineering Task Force, Jun. 2009. Work in progress.

[ID-roll-metrics] Vasseur, J. and Networks, D. Routing Metrics used for Path Calculation in Low Power and Lossy Networks. Internet-Draft draft-ietf-roll-routing-metrics-00, Internet Engineering Task Force, Apr. 2009. Work in progress.

[ID-roll-security] Tsao, T., Alexander, R., Dohler, M., Daza, V. and Lozano, A. A Security Framework for Routing over Low Power and Lossy Networks. Internet-Draft draft-tsao-roll-security-framework-00, Internet Engineering Task Force, Feb. 2009. Work in progress.

[ID-roll-survey] Tavakoli, A., Dawson-Haggerty, S. and Levis, P. Overview of Existing Routing Protocols for Low Power and Lossy Networks. Internet-Draft draft-ietf-roll-protocols-survey-07, Internet Engineering Task Force, Apr. 2009. Work in progress.

[ID-roll-terminology] Vasseur, J. Terminology in Low power And Lossy Networks. Internet-Draft draft-ietf-roll-terminology-01, Internet Engineering Task Force, May 2009. Work in progress.

[ID-roll-trust] Zahariadis, T., Leligou, H., Karkazis, P., Trakadas, P. and Maniatis, S. A Trust Framework for Low Power and Lossy Networks. Internet-Draft draft-zahariad-roll-trust-framework-00, Internet Engineering Task Force, May 2009. Work in progress.

[ID-snmp-optimizations] Mukhtar, H., Joo, S. and Schoenwaelder, J. SNMP optimizations for 6LoWPAN. Internet-Draft draft-hamid-6lowpan-snmp-optimizations-01, Internet Engineering Task Force, April 2009. Work in progress.

[ID-thubert-sfr] Thubert, P. and Hui, J. LoWPAN simple fragment Recovery. Internet-Draft draft-thubert-6lowpan-simple-fragment-recovery-06, Internet Engineering Task Force, Jun. 2009. Work in progress.

[ID-tolle-cap] Tolle, G. A UDP/IP Adaptation of the ZigBee Application Protocol. Internet-Draft draft-tolle-cap-00, Internet Engineering Task Force, Oct. 2008. Work in progress.

[Idesco] Idesco Oy. http://www.idesco.fi.

[IEC62056] IEC 62056. Electricity metering – Data exchange for meter reading, tariff and load control. Tech. rep., IEC, 2006.

[IEEE] The IEEE 802.15 Working Groups. http://www.ieee802.org/15.

[IEEE802.15.4] IEEE Std 802.15.4^{TM}-2006: Wireless Medium Access Control (MAC) and Physical Layer (PHY) Specifications for Low-Rate Wireless Personal Area Networks (LR-WPANs), October 2006.

[IEEE802.15.5] IEEE Std 802.15.5^{TM}-2009: Mesh Topology Capability in Wireless Personal Area Networks (WPANs), May 2009.

[IP500] The IP500 Alliance. http://www.ip500.de.

[IPSO] The IPSO Alliance. http://www.ipso-alliance.org.

[IPSO-Stacks] Abeillé, J., Durvy, M., Hui, J. and Dawson-Haggerty, S. Lightweight IPv6 Stacks for Smart Objects: the Experience of Three Independent and Interoperable Implementations. Tech. Rep. WP 2, IPSO Alliance, Nov 2008.

[ISA100] ISA100, Wireless Systems for Automation. http://www.isa.org/community, May 2008.

[ISA100.11a] ISA100.11a Standard. Wireless Systems for Industrial Automation: Process Control and Related Applications. Tech. rep., ANSI/ISA, April 2009.

[Jennic] Jennic 6LoWPAN. http://www.jennic.com/products.

[Kent87] Kent, C.A. and Mogul, J.C. Fragmentation considered harmful. In *ACM SIGCOMM*, pp. 390–401. 1987.

[KNX] Konnex Association. http://www.knx.org.

[Koodli07] Koodli, R.S. and Perkins, C.E. *Mobile Inter-Networking with IPv6*. A John Wiley and Sons, Inc., 2007.

[MANET] IETF MANET Working Group. http://tools.ietf.org/wg/manet.

[MQTT] MQ Telemetry Transport. http://mqtt.org.

[MQTT-S] Stanford-Clark, A. and Truong, H.L. MQTT for Sensor Networks (MQTT-S), 2008.

[NETLMM] IETF Network-based Localized Mobility Management Working Group. http://tools.ietf.org/wg/netlmm.

[Nivis] Nivis ISA100.11a. http://www.nivis.com.

[oBIX] oBIX. Open Building Information Exchange v1.0. Tech. Rep. obix-1.0-cs-01, OASIS, 2006.

[REST] Representational State Transfer. http://en.wikipedia.org/wiki/rest.

[RFC0768] Postel, J. User Datagram Protocol. RFC 0768, Internet Engineering Task Force, Aug. 1980.

[RFC0791] Postel, J. Internet Protocol. RFC 0791, Internet Engineering Task Force, Sep. 1981.

[RFC0793] Postel, J. Transmission Control Protocol. RFC 0793, Internet Engineering Task Force, Sep. 1981.

[RFC1042] Postel, J. and Reynolds, J. Standard for the transmission of IP datagrams over IEEE 802 networks. RFC 1042, Internet Engineering Task Force, Feb. 1988.

[RFC1112] Deering, S. Host extensions for IP multicasting. RFC 1112, Internet Engineering Task Force, Aug. 1989.

[RFC1122] Braden, R. Requirements for Internet Hosts – Communication Layers. RFC 1122, Internet Engineering Task Force, Oct. 1989.

[RFC1144] Jacobson, V. Compressing TCP/IP Headers for Low-Speed Serial Links. RFC 1144, Internet Engineering Task Force, Feb. 1990.

[RFC1191] Mogul, J. and Deering, S. Path MTU discovery. RFC 1191, Internet Engineering Task Force, Nov. 1990.

[RFC1618] Simpson, W. PPP over ISDN. RFC 1618, Internet Engineering Task Force, May 1994.

[RFC1661] Simpson, W. The Point-to-Point Protocol (PPP). RFC 1661, Internet Engineering Task Force, Jul. 1994.

[RFC1951] Deutsch, P. DEFLATE Compressed Data Format Specification version 1.3. RFC 1951, Internet Engineering Task Force, May 1996.

[RFC1952] Deutsch, P. GZIP file format specification version 4.3. RFC 1952, Internet Engineering Task Force, May 1996.

[RFC1958] Carpenter, B. Architectural Principles of the Internet. RFC 1958, Internet Engineering Task Force, Jun. 1996.

[RFC1981] McCann, J., Deering, S. and Mogul, J. Path MTU Discovery for IP version 6. RFC 1981, Internet Engineering Task Force, Aug. 1996.

[RFC2119] Bradner, S. Key words for use in RFCs to Indicate Requirement Levels. RFC 2119, Internet Engineering Task Force, Mar. 1997.

[RFC2136] Vixie, P., Thomson, S., Rekhter, Y. and Bound, J. Dynamic Updates in the Domain Name System (DNS UPDATE). RFC 2136, Internet Engineering Task Force, Apr. 1997.

[RFC2327] Handley, M. and Jacobson, V. SDP: Session Description Protocol. RFC 2327, Internet Engineering Task Force, Apr. 1998.

[RFC2328] Moy, J. OSPF Version 2. RFC 2328, Internet Engineering Task Force, Apr. 1998.

[RFC2409] Harkins, D. and Carrel, D. The Internet Key Exchange (IKE). RFC 2409, Internet Engineering Task Force, Nov. 1998.

[RFC2460] Deering, S. and Hinden, R. Internet Protocol, Version 6 (IPv6) Specification. RFC 2460, Internet Engineering Task Force, Dec. 1998.

[RFC2464] Crawford, M. Transmission of IPv6 Packets over Ethernet Networks. RFC 2464, Internet Engineering Task Force, Dec. 1998.

[RFC2474] Nichols, K., Blake, S., Baker, F. and Black, D. Definition of the Differentiated Services Field (DS Field) in the IPv4 and IPv6 Headers. RFC 2474, Internet Engineering Task Force, Dec. 1998.

[RFC2507] Degermark, M., Nordgren, B. and Pink, S. IP Header Compression. RFC 2507, Internet Engineering Task Force, Feb. 1999.

[RFC2508] Casner, S. and Jacobson, V. Compressing IP/UDP/RTP Headers for Low-Speed Serial Links. RFC 2508, Internet Engineering Task Force, Feb. 1999.

[RFC2509] Engan, M., Casner, S. and Bormann, C. IP Header Compression over PPP. RFC 2509, Internet Engineering Task Force, Feb. 1999.

[RFC2516] Mamakos, L., Lidl, K., Evarts, J., Carrel, D., Simone, D. and Wheeler, R. A Method for Transmitting PPP Over Ethernet (PPPoE). RFC 2516, Internet Engineering Task Force, Feb. 1999.

[RFC2608] Guttman, E., Perkins, C., Veizades, J. and Day, M. Service Location Protocol, Version 2. RFC 2608, Internet Engineering Task Force, Jun. 1999.

[RFC2616] Fielding, R., Gettys, J., Mogul, J., Frystyk, H., Masinter, L., Leach, P. and Berners-Lee, T. Hypertext Transfer Protocol – HTTP/1.1. RFC 2616, Internet Engineering Task Force, Jun. 1999.

[RFC2671] Vixie, P. Extension Mechanisms for DNS (EDNS0). RFC 2671, Internet Engineering Task Force, Aug. 1999.

[RFC2675] Borman, D., Deering, S. and Hinden, R. IPv6 Jumbograms. RFC 2675, Internet Engineering Task Force, Aug. 1999.

[RFC2801] Burdett, D. Internet Open Trading Protocol – IOTP Version 1.0. RFC 2801, Internet Engineering Task Force, Apr. 2000.

[RFC2911] Hastings, T., Herriot, R., deBry, R., Isaacson, S. and Powell, P. Internet Printing Protocol/1.1: Model and Semantics. RFC 2911, Internet Engineering Task Force, Sep. 2000.

[RFC2960] Stewart, R., Xie, Q., Morneault, K., Sharp, C., Schwarzbauer, H., Taylor, T., Rytina, I., Kalla, M., Zhang, L. and Paxson, V. Stream Control Transmission Protocol. RFC 2960, Internet Engineering Task Force, Oct. 2000.

[RFC3041] Narten, T. and Draves, R. Privacy Extensions for Stateless Address Autoconfiguration in IPv6. RFC 3041, Internet Engineering Task Force, Jan. 2001.

[RFC3053] Durand, A., Fasano, P., Guardini, I. and Lento, D. IPv6 Tunnel Broker. RFC 3053, Internet Engineering Task Force, Jan. 2001.

[RFC3056] Carpenter, B. and Moore, K. Connection of IPv6 Domains via IPv4 Clouds. RFC 3056, Internet Engineering Task Force, Feb. 2001.

[RFC3095] Bormann, C. *et al.* RObust Header Compression (ROHC): Framework and four profiles: RTP, UDP, ESP, and uncompressed. RFC 3095, Internet Engineering Task Force, Jul. 2001.

[RFC3162] Aboba, B., Zorn, G. and Mitton, D. RADIUS and IPv6. RFC 3162, Internet Engineering Task Force, Aug. 2001.

[RFC3168] Ramakrishnan, K., Floyd, S. and Black, D. The Addition of Explicit Congestion Notification (ECN) to IP. RFC 3168, Internet Engineering Task Force, Sep. 2001.

[RFC3241] Bormann, C. Robust Header Compression (ROHC) over PPP. RFC 3241, Internet Engineering Task Force, Apr. 2002.

[RFC3260] Grossman, D. New Terminology and Clarifications for Diffserv. RFC 3260, Internet Engineering Task Force, Apr. 2002.

[RFC3261] Rosenberg, J., Schulzrinne, H., Camarillo, G., Johnston, A., Peterson, J., Sparks, R., Handley, M. and Schooler, E. SIP: Session Initiation Protocol. RFC 3261, Internet Engineering Task Force, Jun. 2002.

[RFC3306] Haberman, B. and Thaler, D. Unicast-Prefix-based IPv6 Multicast Addresses. RFC 3306, Internet Engineering Task Force, Aug. 2002.

[RFC3315] Droms, R., Bound, J., Volz, B., Lemon, T., Perkins, C. and Carney, M. Dynamic Host Configuration Protocol for IPv6 (DHCPv6). RFC 3315, Internet Engineering Task Force, Jul. 2003.

[RFC3344] Perkins, C. IP Mobility Support for IPv4. RFC 3344, Internet Engineering Task Force, Aug. 2002.

[RFC3411] Harrington, D., Presuhn, R. and Wijnen, B. An Architecture for Describing Simple Network Management Protocol (SNMP) Management Frameworks. RFC 3411, Internet Engineering Task Force, Dec. 2002.

[RFC3418] Presuhn, R. Management Information Base (MIB) for the Simple Network Management Protocol (SNMP). RFC 3418, Internet Engineering Task Force, Dec. 2002.

[RFC3439] Bush, R. and Meyer, D. Some Internet Architectural Guidelines and Philosophy. RFC 3439, Internet Engineering Task Force, Dec. 2002.

[RFC3530] Shepler, S., Callaghan, B., Robinson, D., Thurlow, R., Beame, C., Eisler, M. and Noveck, D. Network File System (NFS) version 4 Protocol. RFC 3530, Internet Engineering Task Force, Apr. 2003.

[RFC3544] Koren, T., Casner, S. and Bormann, C. IP Header Compression over PPP. RFC 3544, Internet Engineering Task Force, Jul. 2003.

[RFC3545] Koren, T., Casner, S., Geevarghese, J., Thompson, B. and Ruddy, P. Enhanced Compressed RTP (CRTP) for Links with High Delay, Packet Loss and Reordering. RFC 3545, Internet Engineering Task Force, Jul. 2003.

[RFC3550] Schulzrinne, H., Casner, S., Frederick, R. and Jacobson, V. RTP: A Transport Protocol for Real-Time Applications. RFC 3550, Internet Engineering Task Force, Jul. 2003.

[RFC3551] Schulzrinne, H. and Casner, S. RTP Profile for Audio and Video Conferences with Minimal Control. RFC 3551, Internet Engineering Task Force, Jul. 2003.

[RFC3552] Rescorla, E. and Korver, B. Guidelines for Writing RFC Text on Security Considerations. RFC 3552, Internet Engineering Task Force, Jul. 2003.

[RFC3561] Perkins, C., Belding-Royer, E. and Das, S. Ad hoc On-Demand Distance Vector (AODV) Routing. RFC 3561, Internet Engineering Task Force, Jul. 2003.

[RFC3587] Hinden, R., Deering, S. and Nordmark, E. IPv6 Global Unicast Address Format. RFC 3587, Internet Engineering Task Force, Aug. 2003.

[RFC3602] Frankel, S., Glenn, R. and Kelly, S. The AES-CBC Cipher Algorithm and Its Use with IPsec. RFC 3602, Internet Engineering Task Force, Sep. 2003.

[RFC3610] Whiting, D., Housley, R. and Ferguson, N. Counter with CBC-MAC (CCM). RFC 3610, Internet Engineering Task Force, Sep. 2003.

[RFC3626] Clausen, T. and Jacquet, P. Optimized Link State Routing Protocol (OLSR). RFC 3626, Internet Engineering Task Force, Oct. 2003.

[RFC3633] Troan, O. and Droms, R. IPv6 Prefix Options for Dynamic Host Configuration Protocol (DHCP) version 6. RFC 3633, Internet Engineering Task Force, Dec. 2003.

[RFC3684] Ogier, R., Templin, F. and Lewis, M. Topology Dissemination Based on Reverse-Path Forwarding (TBRPF). RFC 3684, Internet Engineering Task Force, Feb. 2004.

[RFC3697] Rajahalme, J., Conta, A., Carpenter, B. and Deering, S. IPv6 Flow Label Specification. RFC 3697, Internet Engineering Task Force, Mar. 2004.

[RFC3720] Satran, J., Meth, K., Sapuntzakis, C., Chadalapaka, M. and Zeidner, E. Internet Small Computer Systems Interface (iSCSI). RFC 3720, Internet Engineering Task Force, Apr. 2004.

[RFC3775] Johnson, D., Perkins, C. and Arkko, J. Mobility Support in IPv6. RFC 3775, Internet Engineering Task Force, Jun. 2004.

[RFC3819] Karn, P., Bormann, C., Fairhurst, G., Grossman, D., Ludwig, R., Mahdavi, J., Montenegro, G., Touch, J. and Wood, L. Advice for Internet Subnetwork Designers. RFC 3819, Internet Engineering Task Force, Jul. 2004.

[RFC3843] Jonsson, L.-E. and Pelletier, G. RObust Header Compression (ROHC): A Compression Profile for IP. RFC 3843, Internet Engineering Task Force, Jun. 2004.

[RFC3956] Savola, P. and Haberman, B. Embedding the Rendezvous Point (RP) Address in an IPv6 Multicast Address. RFC 3956, Internet Engineering Task Force, Nov. 2004.

[RFC3963] Devarapalli, V., Wakikawa, R., Petrescu, A. and Thubert, P. Network Mobility (NEMO) Basic Support Protocol. RFC 3963, Internet Engineering Task Force, Jan. 2005.

[RFC4122] Leach, P., Mealling, M. and Salz, R. A Universally Unique IDentifier (UUID) URN Namespace. RFC 4122, Internet Engineering Task Force, Jul. 2005.

[RFC4191] Draves, R. and Thaler, D. Default Router Preferences and More-Specific Routes. RFC 4191, Internet Engineering Task Force, Nov. 2005.

[RFC4193] Hinden, R. and Haberman, B. Unique Local IPv6 Unicast Addresses. RFC 4193, Internet Engineering Task Force, Oct. 2005.

[RFC4213] Nordmark, E. and Gilligan, R. Basic Transition Mechanisms for IPv6 Hosts and Routers. RFC 4213, Internet Engineering Task Force, Oct. 2005.

[RFC4291] Hinden, R. and Deering, S. IP Version 6 Addressing Architecture. RFC 4291, Internet Engineering Task Force, Feb. 2006.

[RFC4301] Kent, S. and Seo, K. Security Architecture for the Internet Protocol. RFC 4301, Internet Engineering Task Force, Dec. 2005.

[RFC4302] Kent, S. IP Authentication Header. RFC 4302, Internet Engineering Task Force, Dec. 2005.

[RFC4303] Kent, S. IP Encapsulating Security Payload (ESP). RFC 4303, Internet Engineering Task Force, Dec. 2005.

[RFC4306] Kaufman, C. Internet Key Exchange (IKEv2) Protocol. RFC 4306, Internet Engineering Task Force, Dec. 2005.

[RFC4309] Housley, R. Using Advanced Encryption Standard (AES) CCM Mode with IPsec Encapsulating Security Payload (ESP). RFC 4309, Internet Engineering Task Force, Dec. 2005.

[RFC4389] Thaler, D., Talwar, M. and Patel, C. Neighbor Discovery Proxies (ND Proxy). RFC 4389, Internet Engineering Task Force, Apr. 2006.

[RFC4429] Moore, N. Optimistic Duplicate Address Detection (DAD) for IPv6. RFC 4429, Internet Engineering Task Force, Apr. 2006.

[RFC4443] Conta, A., Deering, S. and Gupta, M. Internet Control Message Protocol (ICMPv6) for the Internet Protocol Version 6 (IPv6) Specification. RFC 4443, Internet Engineering Task Force, Mar. 2006.

[RFC4489] Park, J.-S., Shin, M.-K. and Kim, H.-J. A Method for Generating Link-Scoped IPv6 Multicast Addresses. RFC 4489, Internet Engineering Task Force, Apr. 2006.

[RFC4728] Johnson, D., Hu, Y. and Maltz, D. The Dynamic Source Routing Protocol (DSR) for Mobile Ad hoc Networks for IPv4. RFC 4728, Internet Engineering Task Force, Feb. 2007.

[RFC4815] Jonsson, L.-E., Sandlund, K., Pelletier, G. and Kremer, P. RObust Header Compression (ROHC): Corrections and Clarifications to RFC 3095. RFC 4815, Internet Engineering Task Force, Feb. 2007.

[RFC4830] Kempf, J. Problem Statement for Network-Based Localized Mobility Management (NETLMM). RFC 4830, Internet Engineering Task Force, Apr. 2007.

[RFC4861] Narten, T., Nordmark, E., Simpson, W. and Soliman, H. Neighbor Discovery for IP version 6 (IPv6). RFC 4861, Internet Engineering Task Force, Sep. 2007.

[RFC4862] Thomson, S., Narten, T. and Jinmei, T. IPv6 Stateless Address Autoconfiguration. RFC 4862, Internet Engineering Task Force, Sep. 2007.

[RFC4919] Kushalnagar, N., Montenegro, G. and Schumacher, C. IPv6 over Low-Power Wireless Personal Area Networks (6LoWPANs): Overview, Assumptions, Problem Statement, and Goals. RFC 4919, Internet Engineering Task Force, Aug. 2007.

[RFC4941] Narten, T., Draves, R. and Krishnan, S. Privacy Extensions for Stateless Address Autoconfiguration in IPv6. RFC 4941, Internet Engineering Task Force, Sep. 2007.

[RFC4944] Montenegro, G., Kushalnagar, N., Hui, J. and Culler, D. Transmission of IPv6 Packets over IEEE 802.15.4 Networks. RFC 4944, Internet Engineering Task Force, Sep. 2007.

[RFC4963] Heffner, J., Mathis, M. and Chandler, B. IPv4 Reassembly Errors at High Data Rates. RFC 4963, Internet Engineering Task Force, Jul. 2007.

[RFC4995] Jonsson, L.-E., Pelletier, G. and Sandlund, K. The RObust Header Compression (ROHC) Framework. RFC 4995, Internet Engineering Task Force, Jul. 2007.

[RFC4996] Pelletier, G., Sandlund, K., Jonsson, L.-E. and West, M. RObust Header Compression (ROHC): A Profile for TCP/IP (ROHC-TCP). RFC 4996, Internet Engineering Task Force, Jul. 2007.

[RFC4997] Finking, R. and Pelletier, G. Formal Notation for RObust Header Compression (ROHC-FN). RFC 4997, Internet Engineering Task Force, Jul. 2007.

[RFC5006] Jeong, J., Park, S., Beloeil, L. and Madanapalli, S. IPv6 Router Advertisement Option for DNS Configuration. RFC 5006, Internet Engineering Task Force, Sep. 2007.

[RFC5072] S.Varada, Haskins, D. and Allen, E. IP Version 6 over PPP. RFC 5072, Internet Engineering Task Force, Sep. 2007.

[RFC5213] Gundavelli, S., Leung, K., Devarapalli, V., Chowdhury, K. and Patil, B. Proxy Mobile IPv6. RFC 5213, Internet Engineering Task Force, Aug. 2008.

[RFC5214] Templin, F., Gleeson, T. and Thaler, D. Intra-Site Automatic Tunnel Addressing Protocol (ISATAP). RFC 5214, Internet Engineering Task Force, Mar. 2008.

[RFC5225] Pelletier, G. and Sandlund, K. RObust Header Compression Version 2 (ROHCv2): Profiles for RTP, UDP, IP, ESP and UDP-Lite. RFC 5225, Internet Engineering Task Force, Apr. 2008.

[RFC5444] Clausen, T., Dearlove, C., Dean, J. and Adjih, C. Generalized Mobile Ad hoc Network (MANET) Packet/Message Format. RFC 5444, Internet Engineering Task Force, Feb. 2009.

[RFC5497] Clausen, T. and Dearlove, C. Representing Multi-Value Time in Mobile Ad hoc Networks (MANETs). RFC 5497, Internet Engineering Task Force, Mar. 2009.

[RFC5548] Dohler, M., Watteyne, T., Winter, T. and Barthel, D. Routing Requirements for Urban Low-Power and Lossy Networks. RFC 5548, Internet Engineering Task Force, May 2009.

[ROLL] IETF ROLL Working Group. http://tools.ietf.org/wg/roll.

[SENSEI] EU FP7 SENSEI Project. http://www.sensei-project.eu.

[Sensinode] Sensinode Oy. http://www.sensinode.com.

[SensorML] OGC Sensor Model Language. http://www.opengeospatial.org/standards/sensorml.

[Shan02] Shannon, C., Moore, D. and k claffy Beyond folklore: Observations on fragmented traffic. *IEEE/ACM Transactions on Networking*, 10:709 – 720, 2002.

[Shel03] Shelby, Z., Mähönen, P., Riihijärvi, J.O.R. and Huuskonen, P. NanoIP: The Zen of Embedded Networking. In *Proceedings of the IEEE International Conference on Communications*. May 2003.

[SOAP] Latest SOAP versions. http://www.w3.org/tr/soap.

[SRC81] Saltzer, J., Reed, D. and Clark, D. End-to-end arguments in system design. In *Second International Conference on Distributed Computing Systems*, pp. 509–512. 1981.

[Stevens03] Stevens, W.R., Fenner, B. and Rudoff, A.M. *UNIX Network Programming Volume 1: The Sockets Networking API*, vol. 1. Addison Wesley, 3rd ed., 2003.

[TinyOS] The TinyOS Community. http://www.tinyos.net.

[TinySIP] Krishnamurthy, S. TinySIP: Providing Seamless Access to Sensor-based Services. In *Proceedings of the 1st International Workshop on Advances in Sensor Networks 2006 (IWASN 2006)*. July 2006.

[UPnP] UPnP Forum Specifications. http://upnp.org/standardizeddcps.

[Watteco] Power line communications, W. http://www.watteco.com.

[WBXML] Open Mobile Alliance. Binary XML Content Format Specification. Tech. rep., WAP-192-WBXML-20010725-a, 2001.

[WS] Web Services Architecture. http://www.w3.org/tr/ws-arch.

[WSDL] Web Services Description Language (WSDL) Version 2.0. http://www.w3.org/tr/wsdl20-primer.

[ZigBee] The ZigBee Alliance. http://www.zigbee.org.

[ZigBeeCL] ZigBee Alliance. ZigBee Cluster Library Specification. Tech. rep., ZigBee, October 2007.

[ZigBeeHA] ZigBee Alliance. ZigBee Home Automation Public Application Profile. Tech. rep., ZigBee, October 2007.

[ZigBeeSE] ZigBee Alliance. ZigBee Smart Energy Profile Specification. Tech. rep., ZigBee, May 2008.

Index

6LoWPAN, 3, 6, 7
    addressing, 19
    architecture, 13
    bootstrapping, 20
    header, 20
    link requirements, 17
    link-layer, 17
    mesh, 22
    Neighbor Discovery, 21
    PDU, 31
    protocol stack, 16
    routing, 22
6in4, 122
6to4, 122

61616, 45, 51

AAL5, 55
Access Point, 65
Acknowledgment, 31, 32, 42, 54
Adaptation layer, 27
Address, 31, 34
    EUI-64, 34
    final destination, 39
    link-local, 46
    multicast, 60
    originator, 39
    short, 31, 34, 36
    universal, 35
AES, 86
AH, 87
Alignment, 33, 51
AMI, 146
ANSI C12, 146
Antenna, 84
AODV, 112
Application protocols, 125

compression, 131
design issues, 127
DNS usage, 131
end-to-end, 132
implementation, 156
link issues, 129
network issues, 130
streaming, 132
Applications, 9, 23
Asset management, 92
ATM, 28, 55
Attack, 84
Autoconfiguration, 36
Automatic tunneling, 122

Backhaul link, 82
Backoff, 32
BACnet, 1
Barcode, 65
BC0, 34, 40, 60
Beacon, 30, 192
Binding, 71
Bootstrapping, 20
Border routing, 120
Broadcast, 30, 34, 60
Broker, 137
Building automation, 11, 144

CAP, 140
CBC-MAC, 86
CCM, 86
Channel, 65, 192
Checksum, 42, 45, 47, 50
Chips, 150
Commissioning, 63, 156
Compression
    data, 41

header, 41
    optimistic, 42
Confidentiality, 32
Configuration, 64
Configured tunneling, 122
Context, 42, 46
    default, 47
Contiki, 153, 157
Correspondent node, 97
CSMA/CA, 32

DAD, 69
    optimistic, 188
Datagram, 53, 55
DES, 86
Device identification, 130
DHCP, 67
Diameter, 40
Differentiated services, 47
Dispatch, 32, 34, 47, 56
DLMS, 147
DNS, 53
Door access control, 11
DPWS, 142
DSCP, 47
Dual-stack, 122
Duty-cycle, 5, 93, 131
DYMO, 112

ECN, 48
Edge router, 13, 16, 48, 51
    implementation, 159
EDNS0, 53
Embedded device, 149
Encapsulation, 28, 32
Encryption, 32
End-to-end, 41, 50
Energy industry, 144
Energy reduction, 11
ESP, 87
Ethernet, 27, 52, 65
Ethertype, 28, 32
ETSI, 8
EUI-64, 19, 34, 130

Facility management, 11
FIB, 37

Firewall, 23, 132
Flow, 42
Flow label, 43, 47
Forwarding, 37, 38
Fragment, 41
Fragmentation, 29, 34, 53
Frame, 30
Frame size, 5, 6, 130
Future Internet, 8

GTK, 90

Handover, 93
HC1, 43
HC2, 43, 45
Header compression, 29, 41
Hop Limit, 40, 44, 48
Hop-by-hop, 34, 41
Horizontal networking, 126

IANA, 33
ICMP, 53
ICV, 88
Identification, 54
IEEE 802.15.4, 2, 7, 18
    address, 31
    beacon-enabled, 18, 32, 191
    beaconless, 18, 32, 191
IEEE 802.11, 30, 64, 84
IID, 20, 36, 188
IKE, 87
Implementation, 149
Infrared, 65
Installation, 64
Installer, 65
Integrity, 32
Interface ID, 36, 43
Interface identifier, 188
Internet model, 16
Internet of Things, 1, 3
    size, 3
Internet Protocol, 27
IP500, 8
IPHC, 47, 60
IPsec, 87
IPSO Alliance, 3, 7

IPv4, 52
    interconnectivity, 23, 91, 121
IPv6, 3
    addressing, 20
    interconnectivity, 23
    transition, 121
IPv6-in-IPv4, 122
ISA100, 7, 46

Jumbogram, 52

Key, 32
KNX, 145

Lease, 67
Lollipop, 75
LoWPAN, 13
    ad hoc, 13
    extended, 13

M2M, 1, 8
MAC, 29
Macro-mobility, 93
Maintenance, 11
Management, 5
MANET, 22, 111
Maximum transmission unit, 28, 52
Mesh, 5, 22
Mesh header, 34, 39
Mesh-Under, 22, 38
Micro-mobility, 93
MIPv6, 97
    6LoWPAN problems, 98
    binding, 98
    care-of address, 97
    home address, 97
    Home Agent, 98
    proxy Home Agent, 100
    route optimization, 98
Mobile IP, 97
Mobility, 91, 156
    application layer, 96
    causes, 93
    network, 94
    node, 94
    transport layer, 96
Moore's law, 6

MQTT, 133, 137
MSS, 53
MTU, 23, 28, 52, 54, 186
    path, 53
Multicast, 5, 37, 49, 59

NALP, 33, 35
NanoIP, 7
NAT, 23
NC, 69
Near-field, 65
Neighbor Discovery, 13, 21, 46, 60, 107
    proxy, 108
Neighborhood, 30
NEMO, 102
    prefix delegation, 103
NesC, 153
Network processor, 151
Next header, 34, 44, 54
NHC, 48, 49
Node confirmation, 69
Node registration, 69
NR, 69
NUD, 80, 187

oBIX, 2, 145
OII, 71
OLSR, 114
Optimistic DAD, 188
Optimization, 40, 48, 60
Option, 185
OUI, 34
Overlap, 58

Path MTU detection, 53
Payload Length, 44, 47
PDU, 31, 34, 47
Persistence, 31
PILC, 29
PMIPv6, 100
    local mobility anchor, 101
    mobile access gateway, 101
PMK, 89
PMTUD, 53, 54
Point-to-point, 28
Port, 45, 51, 127
Power line communications, 19

PPP, 27
Prefix, 36
Privacy, 36
Protocol stack, 152
Protosocket, 153, 157
Protothread, 153, 157
Provisioning, 64
Pseudo-header, 47, 50, 185
PSK, 90
PTK, 89
Pub/sub, *see* Publish/subscribe
Publish/subscribe, 133, 137

RA, 185
Real-time, 132
Reassembly, 29, 41, 53, 57
Receiver
    unintended, 50
Redundancy, 43
Registration, 69
Registry, 33
Reliability, 6, 156
REST, 134
RFID, 84
RIB, 37
Roaming, 93
Robustness, 42
ROHC, 27, 41
ROLL, 7, 23, 104
    algorithm, 114
    forwarding, 118
    requirements, 108
    topology, 117
Route redistribution, 120
Route-Over, 22, 40
Router Advertisement, 185
Router on a stick, 38
Routing, 37, 91
    challenges, 104
    distance-vector, 106
    intra-LoWPAN, 105
    link-state, 106
    proactive, 107
    reactive, 107
    requirements, 105, 108
    route metric, 109

Routing protocol, 104
RTCP, 133, 143
RTP, 41, 133, 143

SAA, 20, 36, 67, 188
SAR, 55
Security, 5, 23, 156
    application layer, 131
    firewall, 132
Security association, 87
SENSEI, 8, 11
Service discovery, 141
Short address, 31
Single-chip, 150
SIP, 42, 133, 143
SLP, 141
Smart metering, 11
Smart object, 1
SNMP, 142
SOAP, 134
Socket, 126
Socket API, 127
SPI, 87
SSID, 64
SSLP, 141
Stateless, 36, 42
Stub network, 13
Sub-GHz, 19
Subnetwork, 27
System-on-a-chip, 149, 150

Tag
    datagram, 55
TCP, 53, 126
TID, 75
TinyOS, 153
TinySIP, 144
Traceroute, 40
Tracking, 11
Traffic class, 43, 47
Transaction ID, 75
Two-chip, 151

UDP, 20, 43, 45, 50, 126
uIP, 7, 153, 157
Unslotted, 32

UPnP, 141
USB, 65

Van Jacobson, 41
Version, 47
Visited network, 97
Voice over IP, 42

Web services, 5, 134, 135
    compression, 136
    gateway, 136

Whiteboard, 67, 69
Wireless Embedded Internet, 2, 3
WLAN, 30
WPA, 64
WSDL, 134
WSN, 9

ZigBee, 8, 139
    application layer, 139
    cluster library, 139
    device profile, 140

Printed and bound by CPI Group (UK) Ltd, Croydon, CR0 4YY

27/10/2024

14580217-0003